SpringerBriefs in Physics

For further volumes:
http://www.springer.com/series/8902

Douglas L. Hemmick · Asif M. Shakur

Bell's Theorem and Quantum Realism

Reassessment in Light of the Schrödinger Paradox

 Springer

Douglas L. Hemmick
Boston Drive 66
Berlin, MD 21811
USA
e-mail: douglas196482@yahoo.com

Asif M. Shakur
Department of Physics
Salisbury University
Camden Avenue 1101
Salisbury, MD 21801
USA
e-mail: doc.shaker@gmail.com

ISSN 2191-5423
ISBN 978-3-642-23467-5
DOI 10.1007/978-3-642-23468-2
Springer Heidelberg Dordrecht London New York

e-ISSN 2191-5431
e-ISBN 978-3-642-23468-2

Library of Congress Control Number: 2011937433

Cover design: eStudio Calamar, Berlin/Figueres

Printed on acid-free paper

Springer is part of Springer Science+Business Media (www.springer.com)

Attention has recently been called to the obvious but very disconcerting fact that even though we restrict the disentangling measurements to one *system, the representative obtained for the* other *system is by no means independent of the particular choice of observations which we select for that purpose and which by the way are* entirely *arbitrary.*

Erwin Schrödinger [1] emphasis due to Schrödinger

[1] Schrödinger, E., Proc. Cambridge Phil. Soc. **31**, 555 (1935)

One of us (DH) would like to dedicate this work to all his students at the Osher Lifelong Learning Institute in Lewes, Delaware (formerly SDALL). The other (AS) dedicates it to his wife Lucie and daughter Sophia, in gratitude for their love, support and encouragement.

Acknowledgments

We must thank Sheldon Goldstein of the Rutgers University Department of Mathematics. The authors are extremely grateful to Shelly for providing inspiration and valuable advice. In addition, Travis Norsen assisted by reading through the manuscript and graciously offering his insightful suggestions. We thank also Angela Lahee and Tobias Schwaibold, our editors at Springer. Their enthusiasm and encouragement have been very helpful to us at various stages along the path.

Contents

Chapter 1
Introduction

1.1 Quantum Realism and Bell's Theorem

1.1.1 Opening Remarks

Whatever position one takes on the subject, quantum theory is certainly surprising in its radical break from other fields of physics. Not only does it exhibit indeterminism, but the theory entails an essential denial of objectivity, an abandonment of realism. The latter issue stems from the fact that quantum theory offers little description of physical systems apart from what takes place during measurement processes.[1] Some might point to the fact that the formalism includes the observables, which are supposed to represent physical properties. However, as John S. Bell has stated [1]:

> The concept of 'observable' lends itself to very precise *mathematics* when identified with 'self-adjoint operator'. But physically, it is a rather woolly concept. It is not easy to identify which physical processes are to be given the status of 'observations' and which are to be relegated to the limbo between one observation and another [emphasis due to Bell]

If, in fact the best one can do is to discuss the *responses* of physical entities to certain measurement procedures, then quantum theory seems to imply that the reality of any physical entity depends upon human scrutiny. This is what is meant by a lack of realism or objectivity in the theory.

---

[1] Even the description of measurement lacks clarity, as it depends upon vaguely defined notions of 'system' and 'apparatus.' We submit that as theoretical physicists, it is our duty to seek out sharp, objective definitions of all physical concepts that come into play.

D. L. Hemmick and A. M. Shakur, *Bell's Theorem and Quantum Realism*,
SpringerBriefs in Physics, DOI: 10.1007/978-3-642-23468-2_1,
© The Author(s) 2012

This flaw in quantum theory is something which Albert Einstein disliked,[2] as may be seen in his writings [3, p. 667]:

> What does not satisfy me in that theory, from the standpoint of principle, is its attitude toward that which appears to be the programmatic aim of all physics: the complete description of any (individual) real situation (as it supposedly exists irrespective of any act of observation or substantiation).

With all this, it becomes clear that the search for a realistic theory of quantum phenomena is an issue that might attract great interest. When such a possibility is raised, one topic that comes to the fore is the so-called "no-hidden-variables" theorems, arguments sometimes seen as offering telling evidence against realism, but whose true significance remains controversial. Among these we have not only the theorem of John von Neumann [4][3] but also the efforts[4] of Kochen and Specker [6], and of course Bell's Theorem [7]. Much more recently, a work has been presented by Conway and Kochen in an analysis they call the "Free Will theorem" [8, 9].

The aim of this book[5] is to delve as deeply as possible into the relationship between Bell's Theorem and the possibility of realism in quantum physics, by use of some novel concepts and results. While Bell's Theorem is certainly not new, there is not yet universal accord regarding the ultimate lessons to be drawn. Bell himself and some others regard the result as evidence of a conflict between quantum physics and locality[6] On this view, Bell's Theorem cannot be seen as affecting only quantum realism, but in fact constrains *all* viable interpretations in the same manner. Others[7] feel

---

[2] Some may regard Einstein's position as being summed up by the quotation "God does not play dice." However, this statement was made quite early, and he later became concerned with other difficulties of the theory. In a letter (see [2, p. 221]) to Max Born, Pauli writes: " . . . I was unable to recognize Einstein whenever you talked about him either in your letter or your manuscript. It seemed to me as if you had erected some dummy Einstein for yourself, which you then knocked down with great pomp. In particular Einstein does not consider the concept of 'determinism' to be as fundamental as it is frequently held to be (as he told me emphatically many times) ... he **disputes** that he uses as a criterion for the admissibility of a theory the question 'Is it rigorously deterministic?' "

[3] The page numbers given in citations of this work always refer to the English translation. Chaps. 4 and 6 of the book are reprinted in Wheeler and Zurek's collection [Wheeler, J.A., Zurek, W.H. (eds.) Quantum Theory and Measurement. Princeton University Press, Princeton (1983)] on pages 549–647, although the theorem of von Neumann is not.

[4] As will be discussed in Chap. 2, Kochen and Specker's theorem is closely related to yet another mathematical result that relates to the quantum observables in a similar way. This is Gleason's theorem [5].

[5] This work is based on a doctoral dissertation by Douglas Hemmick. See [10].

[6] See Bell [11, 12], Dürr et al. [13, (Sect. 8)], Maudlin [14, 15], Norsen [16–18] and Wiseman [19].

[7] For the view that the theorem does not constitute a proof of nonlocality see Jarrett [20] and also Evans et al. [21]. Some proceed by citing "counterfactual definiteness," as an attack on the premises that go into the EPR analysis. This position is set forth for example, by Stapp [22] and by Redhead [23]. For more on counterfactuals, see Maudlin [15]. Another issue is what we call the possibility of a "superdeterministic" interpretation (see Bell [24]). This will be discussed in Sect. 3.4.3. One of the most complete sources on all issues surrounding quantum nonlocality is the book by Maudlin [14].

differently, and take Bell's famous result as a serious blow to the possibility of objective quantum formulations. In this work, we shall argue for Bell's position on the issue.

The conclusion we reach regarding Bell's Theorem may not be new, but we shall illustrate the point using original results. In our efforts, we proceed in a twofold manner. First, we review the no-hidden-variables theorems of von Neumann and of Kochen and Specker to provide a clearer background in terms of which to comprehend Bell's Theorem. The second vanguard of our effort will be to present and discuss what we call the Schrödinger paradox. Bell's Theorem, as everyone is aware, is based upon the Einstein–Podolsky–Rosen paradox, and to fully grasp the former depends critically on one's comprehension of the latter. Shortly after the Einstein, Podolsky and Rosen paper appeared, Erwin Schrödinger showed[8] that their result may be generalized so that it applies not just to position and momentum observables, but to *any* observable of the two-particle systems. We will demonstrate here that just as the Einstein–Podolsky–Rosen paradox paves the way for Bell's Theorem and an instance of quantum nonlocality, so what we call the Schrödinger paradox may be combined with any one of a family of theorems, such that a similar result is obtained. In this way, it will be clear that Schrödinger's efforts offer us a broad spectrum of new quantum nonlocality results.

Having examined these matters, the true consequence of Bell's famous theorem then may stand out in sharp relief: that the conflict between quantum theory and locality is not a matter of any theorist's preferences, but of logical inevitability. It should then be clear that the theorem[9] does not serve to limit the possibility of quantum realism as such, but merely requires realistic theories exhibit the same feature every other interpretation must show.

In addition, our efforts will grant us a clear view of the physical scenario under scrutiny in Conway and Kochen's Free Will Theorem. We will see that while their work is correct so far as mathematics is concerned, the conclusion Conway and Kochen draw regarding quantum physics does not follow.[10] It will become clear that their argument is essentially a special case of the Schrödinger paradox, and as such offers another proof of quantum nonlocality rather than leading to the impossibility of determinism in quantum physics.

1.1.2 David Bohm's Theory of Hidden Variables

This discussion would certainly be incomplete without mention of the fact that realism in quantum physics is more than just conjectural. Louis de Broglie's 1926 'Pilot Wave' theory[11] is, in fact, a viable interpretation of quantum phenomena which offers

---

[8] Please see Schrödinger in [25–27].

[9] Readers should also have been reassured that none of the other "no-hidden-variables theorems" imply any true constraint upon realistic quantum as such, but serve only to draw attention to features of quantum theory itself.

[10] See also Bassi and Ghiradi [28], Tumulka [29] and also Goldstein et al. in [30].

[11] See [31–33]. A summary of the early development of the theory is given by de Broglie [34].

an objective picture of the quantum phenomena. When the theory was more systematically presented by David Bohm in 1952,[12] this encouraged de Broglie himself to take up the idea again. We shall refer to the theory as "Bohmian mechanics," after David Bohm.

Besides the feature of objectivity, another reason the success of Bohmian mechanics is important is that it restores *determinism*. Within this theory one may predict, from the present state of the system, its form at any subsequent time. In particular, the results of quantum measurements are so determined from the present state. Determinism is an aspect with which the hidden variables program has been traditionally associated.[13]

The importance of the existence of a successful theory of hidden variables in the form of Bohmian mechanics is perhaps expressed most succinctly by Bell [41]:

> [deBroglie–Bohm quantum theory] does not require, in its very formulation, a vague division of the world into "system" and "apparatus" nor of history into "measurement" and "nonmeasurement." So it applies to the world at large, and not just to idealized laboratory procedures. Indeed the deBroglie–Bohm theory is sharp where the usual one is fuzzy, and general where the usual one is special.

While John S. Bell exhibited clear enthusiasm[14] for Bohmian mechanics, such support is certainly not universal. One reason some have for dismissing the theory seems to be its nonlocal character. However, it is an essential part of this book to convince the reader that conflict with locality is not unique to realistic hidden variables, but is embedded within quantum theory itself.

Once again, we turn to Bell [41]

> That the [Bohmian] guiding wave, in the general case, propagates not in ordinary three-space but in a multi-dimensional configuration space is the origin of the notorious "nonlocality" of quantum mechanics. It is a merit of the deBroglie–Bohm version to bring this out so explicitly that it cannot be ignored.

1.2 Topics to be Covered

Our exposition begins with the mathematical theorems most intimately connected[15] with what is known as "contextuality," namely the analyses of von Neumann, Gleason

---

[12]   See [35]. See also Bell [36, 37] Books on Bohmian mechanics include Holland's 1993 work [38] also Dürr and Teufel's more recent effort published in 2009 [39]. A book which addresses the theory and its general relationship to foundations of quantum mechanics is the 1996 Cushing et al. [40].

[13]   This is the motivation mentioned by von Neumann in his no-hidden variables proof. See Ref. [4].

[14]   For his part, Bell was also intrigued by the theory put forth by Ghirardi et al. [42]. See Bell [43, 44].

[15]   Contextuality, as we will see, is associated with the fact that measurement procedures for each observable are often not unique. Hidden variable theories must take account of the possibility that different measurements are possible for a single observable, and so we must not expect a "one-to-one" mapping from observables to values.

and that of Kochen and Specker.[16] We will review and elaborate on the reasons why these theorems do not disprove the possibility of quantum realism.[17] We will show also that these theorems imply a result we call "spectral-incompatibility". Regarded as proofs of this result, the implications of these theorems toward hidden variables will become more transparent. Our treatment of contextuality concludes as we turn to David Albert's examination of the classic Stern–Gerlach experiment [45]. In considering Albert's example,[18] the implications of contextuality towards the observables can be seen in a new and enlightening way. We will find that the role the quantum theory ascribes to the Hermitian operators, as being direct representatives of physical properties, cannot be taken as literal truth in every situation.[19]

After contextuality, we then turn to the central issue of this book, namely Bell's Theorem and its relationship to quantum nonlocality and quantum realism. While we wish to address this through Erwin Schrödinger's generalization of Einstein–Podolsky–Rosen paradox, nevertheless, we feel it is very worthwhile to first review both the Einstein–Podolsky–Rosen paradox and Bell's Theorem. As we do so, we take care to make explicit the logical structure of each analysis, and the precise role played by realistic hidden variables in each.

Next we present what is perhaps the most distinctive feature of the book, as we focus on the 'Schrödinger paradox,' Erwin Schrödinger's generalization of the EPR paradox. First, we show that the conclusions of Schrödinger's analysis may be reached using a more transparent method which allows one to relate the form of the quantum state to the perfect correlations it exhibits. Second, we use Schrödinger's paradox to derive a wide range of new instances of quantum nonlocality. Like the proof given by Greenberger et al. [48]—these 'Schrödinger nonlocality' proofs are of a deterministic character, i.e., they are 'nonlocality without inequalities' proofs.[20] However, the Schrödinger proofs differ in important ways from GHZ's result, as we shall see.[21]

---

[16] To be precise, von Neumann's theorem falls into a different category than the others. However, examining this classic argument offers an excellent introduction which will be quite useful when we proceed to the more advanced theorems in Chap. 2.

[17] This discussion allows us to understand contextual and noncontextual hidden variables. This will be of great utility not only in making clear the Einstein–Podolsky–Rosen paradox and Bell's Theorem, but we will also find these concepts playing an important role in the discussion of the nonlocality theorems arising from the Schrödinger paradox.

[18] See also Ghirardi [46, pp. 213–217].

[19] A work by Daumer et al. [47] argues for the same conclusion.

[20] See Greenberger et al. [49], and also Mermin [50].

[21] Arguments falling into the same category as the Schrödinger proofs have also been found by Brown and Svetlichny [51] and by Heywood and Redhead [52]. More recent such results were found by Aravind [53], and by Cabello [54].

1.3 Review of the Formalism of Quantum Mechanics

1.3.1 The State and its Evolution

The formalism of a physical theory typically consists of two parts. The first describes the representation of the state of the system; the second, the time evolution of the state. The quantum formalism contains besides these a prescription[22] for the results of measurement.

We first discuss the quantum formalism's representation of state. The quantum formalism associates with every system a Hilbert space $\mathcal{H}$ and represents the state of the system by a vector ψ in that Hilbert space. The vector ψ is referred to as the *wave function*. This is in contrast to the classical description of state which for a system of particles is represented by the coordinates $\{q_1, q_2, q_3, \ldots\}$ and momenta $\{p_1, p_2, p_3, \ldots\}$. We shall often use the Dirac notation: a Hilbert space vector may be written as $|\psi\rangle$, and the inner product of two vectors $|\psi_1\rangle$, $|\psi_2\rangle$ as $\langle\psi_1|\psi_2\rangle$. The quantum mechanical state ψ in the case of a non-relativistic N-particle system is given[23] by a complex-valued function on configuration space, $\psi(\mathbf{q})$, which is an element of L_2—the Hilbert space of square-integrable functions. Here $\mathbf{q}$ represents a point $\{q_1, q_2, \ldots\}$ in the configuration space $\mathbb{R}^{3N}$. The fact that ψ is an element of L_2 implies

$$\langle\psi|\psi\rangle = ||\psi||^2 = \int \mathbf{dq}\,\psi^*(\mathbf{q})\psi(\mathbf{q}) < \infty \tag{1.1}$$

where $\mathbf{dq} = (dq_1 dq_2 \ldots)$. The physical state is defined only up to a multiplicative constant, i.e., $c|\psi\rangle$ and $|\psi\rangle$ (where $c \neq 0$) represent the same state. We may therefore choose to normalize ψ so that (1.1) becomes

$$\langle\psi|\psi\rangle = 1. \tag{1.2}$$

The time evolution of the quantum mechanical wave function is governed by Schrödinger's equation:

$$i\hbar\frac{\partial\psi}{\partial t} = H\psi, \tag{1.3}$$

where H is an operator whose form depends on the nature of the system and in particular on whether or not it is relativistic. To indicate its dependence on time, we write the wave function as ψ_t. For the case of a non-relativistic system, and in the absence of spin, the Hamiltonian takes the form

---

[22] It is the existence of such rules in the quantum mechanical formalism that marks its departure from an objective theory, as we discussed above.

[23] In the absence of spin.

$$H = -\frac{1}{2}\hbar^2 \sum_j \frac{\nabla_j^2}{m_j} + V(\mathbf{q}) \tag{1.4}$$

where $V(\mathbf{q})$ is the potential energy[24] of the system. With the time evolution so specified, ψ_t may be determined from its form ψ_{t_0} at some previous time.

1.3.2 Rules of Measurement

As stated above, the description of the state and its evolution does not constitute the entire quantum formalism. The wave function provides only a formal description and does not by itself make contact with the properties of the system. Using *only* the wave function and its evolution, we cannot make predictions about the typical systems in which we are interested, such as the electrons in an atom, the conduction electrons of a metal, and photons of light. The connection of the wave function to any physical properties is made through the rules of measurement. Because the physical properties in quantum theory are defined through measurement, or observation, they are referred to as 'observables'. The quantum formalism represents the observables by Hermitian operators[25] on the system Hilbert space.

Associated with any observable O is a set of *eigenvectors* and *eigenvalues*. These quantities are defined by the relationship

$$O|\phi\rangle = \mu|\phi\rangle, \tag{1.5}$$

where μ is a real constant. We refer to $|\phi\rangle$ as the eigenvector corresponding to the eigenvalue μ. We label the set of eigenvalues so defined as $\{\mu_a\}$. To each member of this set, there corresponds a set of eigenvectors, all of which are elements of a *subspace* of $\mathcal{H}$, sometimes referred to as the *eigenspace* belonging to μ_a. We label this subspace as $\mathcal{H}_a$. Any two such distinct subspaces are orthogonal, i.e., every vector of $\mathcal{H}_a$ is orthogonal to every vector of $\mathcal{H}_b$ if $\mu_a \neq \mu_b$.

It is important to develop explicitly the example of a *commuting set* of observables $(O^1, O^2, \ldots)$ i.e., where each pair (O^i, O^j) of observables has commutator zero:

$$[O^i, O^j] = O^i O^j - O^j O^i = 0 \tag{1.6}$$

For this case, we have a series of relationships of the form (1.5)

$$O^i|\phi\rangle = \mu^i|\phi\rangle$$
$$i = 1, 2, \ldots \tag{1.7}$$

defining the set of *simultaneous* eigenvalues and eigenvectors. One may also refer to these as *joint-eigenvalues* and *joint-eigenvectors*. Note that (1.5) may be regarded

---

[24] We ignore here the possibility of an external magnetic field.

[25] In the following description, we use the terms observable and operator interchangeably.

as a vector representation of the set of equations (1.7) where $O = (O^1, O^2, \ldots)$ and $\mu = (\mu^1, \mu^2, \ldots)$ are seen respectively as ordered sets of operators and numbers. We use the same symbol, μ, to denote a joint-eigenvalue, as was used to designate an eigenvalue. We emphasize that for a commuting set μ refers to an *ordered set* of numbers. The correspondence of the joint-eigenvalues to the joint-eigenvectors is similar to that between the eigenvalues and eigenvectors discussed above; to each μ_a there corresponds a set of joint-eigenvectors forming a subspace of $\mathcal{H}$ which we refer to as the *joint-eigenspace* belonging to μ_a.

We denote by P_a the projection operator onto the eigenspace belonging to μ_a. It will be useful for our later discussion to express P_a in terms of an orthonormal basis of $\mathcal{H}_a$. If ϕ_k is such a basis, we have

$$P_a = \sum_k |\phi_k\rangle\langle\phi_k|, \tag{1.8}$$

which means that the operator which projects onto a subspace is equal to the sum of the projections onto the one-dimensional spaces defined by any basis of the subspace. If we label the eigenvalues of the observable O as μ_a then O is represented in terms of the P_a by

$$O = \sum_a \mu_a P_a. \tag{1.9}$$

Having developed these quantities, we now discuss the rules of measurement. The first rule concerns the possible outcomes of a measurement and it states that they are restricted to the eigenvalues $\{\mu_a\}$ for the measurement of a single observable[26] O or *joint*-eigenvalues $\{\mu_a\}$ for measurement a commuting set of observables $(O^1, O^2, \ldots)$.

The second rule provides the probability of the measurement result equaling one particular eigenvalue or joint-eigenvalue:

$$P(O = \mu_a) = \langle\psi|P_a|\psi\rangle$$
$$P((O^1, O^2, \ldots) = \mu_a) = \langle\psi|P_a|\psi\rangle \tag{1.10}$$

where the former refers to the measurement of a single observable and the latter to a commuting set. As a consequence of the former, the expectation value for the result of a measurement of O is given by

$$E(O) = \sum_a \mu_a \langle\psi|P_a|\psi\rangle = \langle\psi|O|\psi\rangle \tag{1.11}$$

where the last equality follows from (1.9).

---

[26] A measurement may be classified either as *ideal* or *non-ideal*. Unless specifically stated, it will be assumed whenever we refer to a measurement process, what is said will apply just as well to *either* of these situations.

The third rule governs the effect of measurement on the system's wave function. It is here that the measurement is governed by a different rule depending on whether one performs an ideal or non-ideal measurement. An ideal measurement is defined as one for which the wave function's form after measurement is given by the (normalized) projection of ψ onto the eigenspace $\mathcal{H}_a$ of the measurement result μ_a

$$P_a \psi / \| P_a \psi \|. \tag{1.12}$$

The case of a non-ideal measurement arises when an ideal measurement of a commuting set of observables $(O^1, O^2, \ldots)$ is regarded as a measurement of an *individual* member of the set. An ideal measurement of the *set* of observables leaves the wave function as the projection of ψ onto their *joint*-eigenspace. For an arbitrary vector ψ, the projection onto the eigenspace of an individual member of the set does not generally equal the projection onto the set's joint-eigenspace. Therefore an ideal measurement of a commuting set cannot be an ideal measurement of an individual member of the set. We refer to the procedure as a *non-ideal* measurement of the observable. This completes the presentation of the general rules of measurement. We discuss below two special cases for which these rules reduce to a somewhat simpler form.

The first special case is that of a *non-degenerate* observable. The eigenspaces of such an observable are one-dimensional and are therefore spanned by a single normalized vector called an eigenvector. We label the eigenvector corresponding to eigenvalue μ_a as $|a\rangle$. The operator which projects onto such a one-dimensional space is written as:

$$P_a = |a\rangle\langle a|. \tag{1.13}$$

It is easy to see that the projection given in (1.8) reduces to this form. The form of the observable given in (1.9) then reduces to:

$$O = \sum_a \mu_a |a\rangle\langle a|. \tag{1.14}$$

The probability of a measurement result being equal to μ_a is then

$$P(O = \mu_a) = |\langle a|\psi\rangle|^2, \tag{1.15}$$

in which case the wave function subsequent to the measurement is given by $|a\rangle$. These rules are perhaps the most familiar ones since the non-degenerate case is usually introduced first in presentations of quantum theory.

Finally, there is a case for which the measurement rules are essentially the same as these, and this is that of a set of commuting observables which form a *complete set*. The joint-eigenspaces of a complete set are one-dimensional. The rules governing the probability of the measurement result μ_i and the effect of measurement on the wave function are perfectly analogous to those of the non-degenerate observable. This completes our discussion of the rules of measurement.

1.4 Von Neumann's Theorem and Hidden Variables

1.4.1 Introduction

The quantum formalism contains features which may be considered objectionable by some. These are its lack of realism and indeterminism. The aim of the development of a hidden variables theory[27] is to give a formalism which, while being empirically equivalent to the quantum formalism, does not possess these features. In the present section we shall present and discuss one of the earliest works to address the hidden variables question, which is the 1932 analysis of John von Neumann.[28] We shall also review and elaborate on Bell's analysis [55] of this work, in which he made clear its limitations.

Von Neumann's hidden variables analysis appeared in his now classic book *Mathematical Foundations of Quantum Mechanics*. This book is notable both for its exposition of the mathematical structure of quantum theory, and as one of the earliest works[29] to systematically address both the hidden variables issue and the measurement problem. The quantum formalism presents us with two different types of state function evolution: that given by the Schrödinger equation and that which occurs during a measurement. The latter evolution appears in the formal rule given above in Eq. 1.12. The measurement problem is the problem of the reconciliation of these two types of evolution.

In his analysis of the hidden variables problem, von Neumann proved a mathematical result now known as von Neumann's Theorem and then argued that this theorem implied the very strong conclusion that no hidden variables theory can provide empirical agreement with quantum mechanics: (preface pp. ix, x) " . . . such an explanation (by 'hidden parameters') is incompatible with certain qualitative fundamental postulates of quantum mechanics." The author further states [4, p. 324]: "It should be noted that we need not go further into the mechanism of the 'hidden parameters,' since we now know that the established results of quantum mechanics can never be

---

[27] See the discussion of Bohmian mechanics above. The success of this theory is that it explains quantum phenomena without such features. The general issue of hidden variables is discussed in Bell [36, 55, 56]. Many of the works by Bell which are of concern to us may be found in [Bell, J.S.: Speakable and Unspeakable in Quantum Mechanics, Cambridge University Press, Cambridge (1987)]. See also [Bell, M., Gottfried, K., Veltman, M., John S. Bell on the Foundations of Quantum Mechanics. World Scientific Publishing Company, Singapore (2001)] and [Bell, M., Gottfried, K., Veltman, M., (eds.) Quantum Mechanics, High Energy Physics and Accelerators: Selected Papers of John S. Bell (with commentary). World Scientific Publishing Company, Singapore (1995)]. The latter two are complete collections containing all of Bell's papers on quantum foundations. See also Bohm [35, 58], Belinfante [59], Hughes [60] and Jammer [61].

[28] The original work is [4]. Discussions of von Neumann's hidden variables analysis may be found within [36, 55, 62, 63]

[29] Within Chap. 4 of the present work, we shall discuss a 1935 analysis by Erwin Schrödinger [25]. This is the paper in which the 'Schrödinger's cat' paradox first appeared, but it contains also other significant results such as Schrödinger's generalization of the Einstein–Podolsky–Rosen paradox.

re-derived with their help." The first concrete demonstration[30] that this claim did not follow was given in 1952 when David Bohm constructed a viable theory of hidden variables [35]. Then in 1966, Bell [55] analyzed von Neumann's argument against hidden variables and showed it to be in error. In this section, we begin by discussing an essential concept of von Neumann's analysis: the state representation of a hidden variables theory. We then present von Neumann's Theorem and no-hidden-variables argument. Finally, we show where the flaw in this argument lies.

The analysis of von Neumann is concerned with the description of the *state* of a system, and the question of the *incompleteness* of the quantum formalism's description. The notion of the incompleteness of the quantum mechanical description was particularly emphasized by Einstein. The Einstein–Podolsky–Rosen paper was designed as a proof of such incompleteness and the authors concluded this work with the following statement [64]: "*While we have thus shown that the wave function does not provide a complete description of the physical reality, we left open the question of whether or not such a description exists. We believe, however, that such a theory is possible.*" The hidden variables program, which is an endeavor to supplement the state description, is appears to be just the type of program Einstein was advocating. A complete state description might be constructed to eliminate the objectionable features of the quantum theoretical description.

The particular issue which von Neumann's analysis addressed was the following: is it possible to restore *determinism* to the description of physical systems by the introduction of hidden variables into the description of the state of a system. The quantum formalism's state representation given by ψ does not generally permit deterministic predictions regarding the values of the physical quantities, i.e. the observables. Thus results obtained from performance of measurements on systems with *identical* state representations ψ may be expected to vary. (The statistical quantity called the "dispersion" is used to describe this variation quantitatively). While it generally does not provide predictions for each individual measurement of an observable O, the quantum formalism does give a prediction for its average or *expectation* value[31]:

$$E(O) = \langle \psi | O | \psi \rangle. \tag{1.17}$$

Von Neumann's analysis addresses the question of whether the lack of determinism in the quantum formalism may be ascribed to the fact that the state description as given by ψ is incomplete. If this were true, then the complete description of state—consisting of both ψ and an additional parameter we call λ—should allow one to make predictions regarding individual measurement results for each observable. Note that such predictability can be expressed mathematically by stating that for every ψ and λ,

---

[30]   See Jammer's book [61, p. 272] for a discussion of an early work by Grete Hermann that addresses the impact of von Neumann's Theorem.

[31]   When generalized to the case of a mixed state, this becomes

$$E(O) = \text{Tr}(UO) \tag{1.16}$$

where U is a positive operator with the property $\text{Tr}(U) = 1$. Here U is known as the "density matrix". See for example, [65, p. 378]

there must exist a "value map" function, i.e. a mathematical function assigning to each observable its value. We represent such a value map function by the expression $V_\lambda^\psi(O)$. Von Neumann referred to a hypothetical state described by the parameters ψ and λ as a "dispersion-free state," since results obtained from measurements on systems with identical state representations in ψ and λ are expected to be identical and therefore exhibit no dispersion.

Von Neumann's theorem is concerned with the general form taken by a function $E(O)$, which assigns to each observable its expectation value. The function addressed by the theorem is considered as being of sufficient generality that the expectation function of quantum theory, or of any empirically equivalent theory must assume the form von Neumann derived. In the case of quantum theory, $E(O)$ should take the form of the quantum expectation formula (1.16). In the case of a dispersion-free state, corresponding to fixed λ, a series of measurements of a given observable O should simply present the same value each time.[32] The expectation value of O, i.e., $E(O)$ should then be the value obtained in each run. Therefore, for a dispersion free state, the function $E(O)$ should be a mapping from the observables to their values.

When one analyzes the form of $E(O)$ developed in the theorem, it is easy to see that it cannot be a function of the latter type. Von Neumann went on to conclude from this that no theory involving dispersion-free states can agree with quantum mechanics. However, since the theorem places an unreasonable restriction on the function $E(O)$, this conclusion does *not* follow.

1.4.2 Von Neumann's Theorem

The assumptions regarding the function $E(O)$ are as follows. First, the value E assigns to the 'identity observable' $\mathbf{1}$ is equal to unity:

$$E(\mathbf{1}) = 1 \tag{1.18}$$

The identity observable is the projection operator associated with the entire system Hilbert space. All vectors are eigenvectors of $\mathbf{1}$ of eigenvalue 1. The second assumption is that E of any real-linear combination of observables is the same linear combination of the values E assigns to each individual observable:

$$E(aA + bB + \cdots) = aE(A) + bE(B) + \cdots \tag{1.19}$$

where $(a, b, ...)$ are real numbers and $(A, B, ...)$ are observables. Finally, it is assumed that E of any projection operator P must be non-negative:

---

[32] Of course, this raises the question of whether it is possible to prepare an ensemble of states with fixed λ. Whether the answer to this question is yes or no, the existence of dispersion free states still implies that it is possible to write a value map function $V_\lambda^\psi(O)$ on the observables. For fixed ψ, it is the variation in λ which is considered responsible for the variation in results of measurement. If, on the other hand, we fix *both ψ and λ* then we are dealing with just one member of the ensemble and there should be just a *single value assigned to each observable*. This is the meaning of the value map $E(O)$ in the case of dispersion free states.

$$E(P) \geq 0. \tag{1.20}$$

For example, in the case of the value map function V_λ^ψ, P must be assigned either 1 or 0, since these are its possible values. According to the theorem, these premises imply that $E(O)$ must be given by the form

$$E(O) = \text{Tr}(UO) \tag{1.21}$$

where U is a positive operator with the property $\text{Tr}(U) = 1$.

The demonstration[33] of this conclusion is straightforward. We begin by noting that any operator O may be written as a sum of Hermitian operators. Define A and B by the relationships $A = \frac{1}{2}(O + O^\dagger)$ and $B = \frac{1}{2i}(O - O^\dagger)$, where $O^\dagger$ is the Hermitian conjugate of O. Then it is easily seen that A and B are Hermitian and that

$$O = A + iB. \tag{1.22}$$

We define the function $E^*(O)$ by

$$E^*(O) = E(A) + iE(B), \tag{1.23}$$

where $E(O)$ is von Neumann's $E(O)$, and A and B are defined as above. From the Eqs. 1.19 and 1.23 we have that $E^*(O)$ has the property of complex linearity. Note that $E^*(O)$ is a generalization of von Neumann's $E(O)$: the latter is a *real*-linear function on the Hermitian operators, while the former is a *complex*-linear function on *all* operators. The general form of E^* will now be investigated for the case of a finite-dimensional operator expressed as a matrix in terms of some orthonormal basis. We write the operator O in the form

$$O = \sum_{m,n} |m\rangle\langle m|O|n\rangle\langle n|, \tag{1.24}$$

where the sums over m, n are finite sums. This form of O is a linear combination of the operators $|m\rangle\langle n|$, so that the complex linearity of E^* implies that

$$E^*(O) = \sum_{m,n} \langle m|O|n\rangle E^*(|m\rangle\langle n|). \tag{1.25}$$

Now we define the operator U by the relationship $U_{nm} = E^*(|m\rangle\langle n|)$ and the Eq. 1.25 becomes

$$E^*(O) = \sum_{m,n} O_{mn}U_{nm} = \sum_m (UO)_{mm} = \text{Tr}(UO). \tag{1.26}$$

Since von Neumann's $E(O)$ is a special case of $E^*(O)$, (1.26) implies that

---

[33] A proof of the theorem may be found in von Neumann's original work [4]. Albertson presented a simplification of this proof in 1961 [62]. What we present here is a further simplification.

$$E(O) = \text{Tr}(UO). \tag{1.27}$$

We now show that U is a positive operator. It is a premise of the theorem that $E(P) \geq 0$ for any projection operator P. Thus we write $E(P_\chi) \geq 0$ where P_χ is a one-dimensional projection operator onto the vector χ. Using the form of E found in (1.27), we have[34]

$$\text{Tr}(U P_\chi) = \langle \chi | U | \chi \rangle \geq 0. \tag{1.28}$$

Since χ is an arbitrary vector, it follows that U is a positive operator. The relation $\text{Tr}(U) = 1$ is shown as follows: from the first assumption of the theorem (1.18) together with the form of E given by (1.27) we have $\text{Tr}(U) = \text{Tr}(U\mathbf{1}) = 1$. This completes the demonstration of von Neumann's Theorem.

1.4.3 Von Neumann's Impossibility Proof

We now present von Neumann's argument against the possibility of hidden variables. Consider the function $E(O)$ evaluated on the one-dimensional projection operators P_ϕ. For such projections, we have the relationship

$$P_\phi = P_\phi^2. \tag{1.29}$$

As mentioned above, in the case when $E(O)$ is to correspond to a dispersion free state represented by some ψ and λ, it must map the observables to their values. We write $V_\lambda^\psi(O)$ for the value map function corresponding to the state specified by ψ and λ. Von Neumann noted $V_\lambda^\psi(O)$ must obey the relation:

$$f(V_\lambda^\psi(O)) = V_\lambda^\psi(f(O)), \tag{1.30}$$

where f is any mathematical function. This is easily seen by noting that the quantity $f(O)$ can be measured by measuring O and evaluating f of the result. This means that the value of the observable $f(O)$ will be f of the value of O. Thus, if $V_\lambda^\psi(O)$ maps each observable to a value then we must have (1.30). Hence $V_\lambda^\psi(P_\phi^2) = (V_\lambda^\psi(P_\phi))^2$ which together with (1.29) implies

$$V_\lambda^\psi(P_\phi) = (V_\lambda^\psi(P_\phi))^2 \tag{1.31}$$

This last relationship implies that $V_\lambda^\psi(P_\phi)$ must be equal to either 0 or 1.

---

[34] The equality $\text{Tr}(U P_\chi) = \langle \chi | U | \chi \rangle$ in (1.28) is seen as follows. The expression $\text{Tr}(U P_\chi)$ is independent of the orthonormal basis ϕ_n in terms of which the matrix representations of U and P_χ are expressed, so that one may choose an orthonormal basis of which $|\chi\rangle$ itself is a member. Since $P_\chi = |\chi\rangle\langle\chi|$, and $P_\chi|\phi_n\rangle = 0$ for all $|\phi_n\rangle$ except $|\chi\rangle$, we have $\text{Tr}(U P_\chi) = \langle \chi | U | \chi \rangle$.

Recall from the previous section the relation $E(P_\phi) = \langle\phi|U|\phi\rangle$. If $E(O)$ takes the form of a value map function such as $V_\lambda^\psi(P_\phi)$, then it follows that the quantity $\langle\phi|U|\phi\rangle$ is equal to either 0 or 1. Consider the way this quantity depends on vector $|\phi\rangle$. If we vary $|\phi\rangle$ continuously then $\langle\phi|U|\phi\rangle$ will also vary continuously. If the only possible values of $\langle\phi|U|\phi\rangle$ are 0 and 1, it follows that this quantity must be a *constant*, i.e. we must have either $\langle\phi|U|\phi\rangle = 0 \ \forall\phi \in \mathcal{H}$, or $\langle\phi|U|\phi\rangle = 1 \forall\phi \in \mathcal{H}$. If the former holds true, then it must be that U itself is zero. However, with this and the form (1.27), we find that $E(1)=0$; a result that conflicts with the theorem assumption that $E(1)=1$ (1.18). Similarly, if $\langle\phi|U|\phi\rangle = 1$ for all $\phi \in \mathcal{H}$ then it follows that $U=1$. This result also conflicts with the requirement (1.18), since it leads to $E(1) = \text{Tr}(1) = n$ where n is the dimensionality of $\mathcal{H}$.

From the result just obtained, one can conclude that any function $E(O)$ which satisfies the constraints of von Neumann's Theorem [see (1.18–1.20)] must fail to satisfy the relationship (1.30), and so cannot be a value map function on the observables.[35] From this result, von Neumann concluded that it is impossible for a deterministic hidden variables theory to provide empirical agreement with quantum theory [4, p. 325]:

> It is therefore not, as is often assumed, a question of a re-interpretation of quantum mechanics—the present system of quantum mechanics would have to be objectively false in order that another description of the elementary processes than the statistical one be possible.

1.4.4 Refutation of the Impossibility Proof

While it is true that the mathematical theorem of von Neumann is a valid one, it is not the case that the impossibility of hidden variables follows. The invalidity of von Neumann's argument against hidden variables was shown by Bohm's development [35] of a successful hidden variables theory (a *counter-example* to von Neumann's proof) and by Bell's explicit analysis [55] of von Neumann's proof. We will now present the latter.

The no hidden variables demonstration of von Neumann may be regarded as consisting of two components: a mathematical theorem and an analysis of its implications toward hidden variables. As we have said, the theorem itself is correct when regarded as pure mathematics. The flaw lies in the analysis connecting this theorem to hidden variables. The conditions prescribed for the function E are found in Eqs. 1.18–1.20. The theorem of von Neumann states that from these assumptions follows the conclusion that the form of $E(O)$ must be given by (1.27). When one considers an actual physical situation, it becomes apparent that the second of the theorem's conditions is not at all a reasonable one. As we shall see, the departure of this condition from

---

[35] It should be noted that the same result may be proven without use of (1.30) since the fact that $V_\lambda^\psi(P_\phi)$ must be either 0 or 1 follows simply from the observation that these are the eigenvalues of P_ϕ.

being a reasonable constraint on $E(O)$ is marked by the case of its application to non-commuting observables.

We wish to demonstrate why the assumption (1.19) is an unjustified constraint on E. To do so, we first examine a particular case in which such a relationship is reasonable, and then contrast this with the case for which it is not. The assumption itself calls for the real-linearity of $E(O)$, i.e. that E must satisfy $E(aA + bB + \cdots) = aE(A) + bE(B) + \cdots$ for any observables $\{A, B, \ldots\}$ and real numbers $\{a, b, \ldots\}$. This is in fact, a sensible requirement for the case where $\{A, B, \ldots\}$ are *commuting* observables. Suppose for example, the observables O_1, O_2, O_3 form a commuting set and that they obey the relationship $O_1 = O_2 + O_3$. We know from the quantum formalism that one may measure these observables simultaneously and that the the the measurement result $\{o_1, o_2, o_3\}$ must be a member of the joint-eigenspectrum of the set. By examining the relation (1.7) which defines the joint-eigenspectrum, it is easily seen that any member of the joint eigenspectrum of O_1, O_2, O_3 must satisfy $o_1 = o_2 + o_3$. This being the case, one might well expect that the function $E(O)$—which in the case of a dispersion free state must be a value map $V_\lambda^\psi(O)$ on the observables—should be required to satisfy $E(O_1) = E(O_2) + E(O_3)$. On the other hand, suppose we consider a set $\{O, P, Q\}$ satisfying $O = P + Q$, where the observables P and Q *fail to commute*, i.e. $[P, Q] \neq 0$. It is easy to see that O commutes with neither P nor Q. It is therefore impossible to perform a simultaneous measurement of any two of these observables. Hence, measurements of these observables require three *distinct* experimental procedures. This being so, there is *no justification* for the requirement that $E(O) = E(P) + E(Q)$ for such cases.

As an example one may consider the case of a spin $\frac{1}{2}$ particle. Suppose that the components of the spin given by σ_x, σ_y and σ' where

$$\sigma' = \frac{1}{\sqrt{2}}(\sigma_x + \sigma_y), \tag{1.32}$$

are to be examined. The measurement procedure for any given component of the spin of a particle is performed by a suitably oriented Stern–Gerlach magnet. For example, to measure the x-component, the magnet must be oriented along the x-axis; for the y-component it must be oriented along the y-axis. A measurement of σ' is done using a Stern–Gerlach magnet oriented along an axis in yet another direction.[36] The relationship (1.32) cannot be a reasonable demand to place on the expectation function $E(O)$ of the observables σ_x, σ_y, σ', since these quantities are measured using completely distinct procedures.

Thus von Neumann's hidden variables argument is seen to be an unsound one. That it is based on an unjustified assumption is sufficient to show this. It should also be noted that the presence of the real-linearity postulate discussed above makes von Neumann's entire case against hidden variables into an argument of a rather trivial character. Examining the above example involving the three spin components of a

---

[36] It is not difficult to show that σ' defined in this way is the spin component along an axis which is in the x, y plane and lies at $45°$ from both the x and y axis.

spin $\frac{1}{2}$ particle, we find that the eigenvalues of these observables $\pm\frac{1}{2}$ do not obey (1.32), i.e.

$$\pm\frac{1}{2} \neq \frac{1}{\sqrt{2}} \left(\pm\frac{1}{2} \pm \frac{1}{2} \right) \tag{1.33}$$

Since $E(O)$ by hypothesis must satisfy (1.32), it cannot map the observables to their eigenvalues. Hence, with the real-linearity assumption one can almost immediately 'disprove' hidden variables. It is therefore apparent that Von Neumann's case against hidden variables rests essentially upon the arbitrary requirement that $E(O)$ obey real linearity—an assumption that gives immediate disagreement with the simple and natural demand that E agree with quantum mechanics in giving the eigenvalues as the results of measurement.

1.4.5 Summary and Further Remarks

In our discussion of von Neumann's no hidden variables argument, we found that the argument may be regarded as consisting of two components: a theorem which concerns the general form for an expectation function $E(O)$ on the observables, and a proof that the function $E(O)$ so developed cannot be a value map function. Because the assumption of the real-linearity of $E(O)$ is an unjustified one, the work of von Neumann does *not* imply the general failure of hidden variables. Finally, we noted that assuming *only* the real-linearity of E, one may easily arrive at the conclusion that such a function cannot be a map to the observables' eigenvalues. Ultimately, the lesson to be learned from von Neumann's theorem is simply that there exists no mathematical function from observables to their values obeying the requirement of real-linearity.

Abner Shimony has reported [66] that Albert Einstein was aware of both the von Neumann analysis itself and the reason it fails as a hidden variables impossibility proof. The source of Shimony's report was a personal communication with Peter Bergmann. Bergmann reported that during a conversation with Einstein regarding the von Neumann proof, Einstein opened von Neumann's book to the page where the proof is given, and pointed to the linearity assumption. He then stated[37] that there is no reason why this premise should hold in a state not acknowledged by quantum mechanics, if the observables are not simultaneously measurable. Here the "state not acknowledged by quantum mechanics" seems to refer to von Neumann's dispersion-free state, i.e. the state specified by ψ, and λ. It is almost certain that Erwin Schrödinger would also have realized the error in von Neumann's impossibility proof. In his 1935 paper [25] Schrödinger gives an analysis quite similar to von

---

[37] Reference to such a remark by Einstein is also found in Gilder's recent book. See [67, pp. 160–161].

Neumann's Theorem, yet does not[38] arrive at the same 'impossibility' conclusion. We shall discuss Schrödinger's derivation in the next section and offer a more detailed presentation in Chap. 4. In view of the scarcity[39] of early responses to von Neumann's proof, it is valuable to have such evidence of Einstein's and Schrödinger's awareness of the argument and its shortcomings.

In our introduction to von Neumann's Theorem, we stated that the existence of a deterministic hidden variables theory led to the result that for each ψ and λ, there exists a value map on the observables. We represented such value maps by the expression $V_\lambda^\psi (O)$. If one considers the question of hidden variables more deeply, it is clear that the agreement of their predictions with those of quantum mechanics requires an additional criterion that beyond the existence of a value map for each ψ and λ: it requires agreement with the *statistical* predictions of the quantum formalism. To make possible the empirical agreement of quantum theory, in which only statistical predictions are generally possible with the deterministic description of a hidden variables theory, we regard their descriptions of a quantum system in the following way. The quantum mechanical state given by ψ corresponds to a *statistical ensemble* of the states given by ψ and λ; the members of the ensemble being described by the same ψ, but differing in λ. The variation in measurement results found for a series of quantum mechanical systems with identical ψ is to be accounted for by the variation in parameter λ among the ensemble of ψ, λ states. For precise agreement in this regard, we require that for all ψ and O, the following relationship must hold:

$$\int_{-\infty}^{\infty} d\lambda \rho(\lambda) V_\lambda^\psi (O) = \langle \psi | O | \psi \rangle, \tag{1.34}$$

where $\rho(\lambda)$ is the probability distribution over λ.

We have seen from von Neumann's result and from our simple examination of the spin $\frac{1}{2}$ observables σ_x, σ_y, $\frac{1}{\sqrt{2}}(\sigma_x + \sigma_y)$, that it is impossible to develop a linear function mapping the observables to their eigenvalues. We shall find also that an impossibility proof may be developed showing that the criterion of agreement with the quantum statistics, i.e., the agreement with (1.34), cannot be met by functions of the form $V_\lambda^\psi (O)$. Bell's Theorem is, in fact, such a proof. We will present the proof of Bell's result along with some further discussion in Chap. 3.

1.4.6 Schrödinger's Derivation of Von Neumann's Proof

As mentioned above, in his famous "cat paradox" paper [25], Schrödinger presented an analysis which, as far as hidden variables are concerned, is essentially equivalent

---

[38]   In fact, if Schrödinger *had* interpreted his result this way, this—in light of his own generalization of the EPR paradox presented in the same paper–would have allowed him to reach a further and very striking conclusion. We shall discuss this in Chap. 4.

[39]   See Max Jammer's book [61, p. 265].

to the von Neumann proof. Schrödinger's study of the problem was motivated by the results of his generalization of the Einstein–Podolsky–Rosen paradox. While EPR had concluded definite values on the position and momentum observables only, Schrödinger was able to show that such values must exist for *all* observables of the state considered by EPR. We will discuss both the original Einstein–Podolsky–Rosen analysis, and Schrödinger's generalization thereof in much more detail in Chap. 4. To probe the possible relationships which might govern the values assigned to the various observables, Schrödinger then gave a brief analysis of a system whose Hamiltonian takes the form

$$H = p^2 + a^2 q^2. \tag{1.35}$$

We are aware from the well-known solution of the harmonic oscillator problem, that this Hamiltonian's eigenvalues are given by $\{a\hbar, 3a\hbar, 5a\hbar, \ldots\}$. Consider a mapping $V(O)$ from observables to values. If we require that the assignments V makes to the observables H, p, q satisfy (1.35) then we must have

$$V(H) = (V(p))^2 + a^2(V(q))^2, \tag{1.36}$$

which implies

$$\left((V(p))^2 + a^2(V(q))^2\right) \Big/ a\hbar = \text{an odd integer}. \tag{1.37}$$

This latter relationship cannot generally be satisfied by the eigenvalues of q, and p—each of which may be any real number—and an arbitrary positive number a.

The connection of this result to the von Neumann argument is immediate. In the discussion of Sect. 1.4.3, we noted that the value of the observable $f(O)$ will be f of the value of O, so that any value map must satisfy $f(V(O)) = V(f(O))$, as given in Eq. 1.30. Here f may be any mathematical function. It follows from this that (1.36) is equivalent to a relation between the observables H, p^2, and q^2 given by:

$$V(H) = V(p^2) + a^2 V(q^2). \tag{1.38}$$

With the known eigenvalues of H, this leads to

$$\left(V(p^2) + a^2 V(q^2)\right) / a\hbar = \text{an odd integer}, \tag{1.39}$$

which cannot generally be satisfied by the eigenvalues of q^2, and p^2—each of which may be any positive real number—and an arbitrary positive number a. We have here another example leading to a demonstration of von Neumann's result that there is no linear value map on the observables (Recall the example of the spin component observables—σ_x, σ_y and $\sigma' = \frac{1}{\sqrt{2}}(\sigma_x + \sigma_y)$ given above). If we consider the von Neumann function $E(O)$, the real-linearity assumption requires it to satisfy (1.38). Therefore, $E(O)$ cannot map the observables to their eigenvalues. Schrödinger did not regard this as proof of the impossibility of hidden variables, as von Neumann did, but concluded only that the relationships such as (1.35) will not necessarily be satisfied by the value assignments made to the observables constrained by such a relation. Indeed, if Schrödinger *had* made von Neumann's error of interpretation, this would contradict results he had developed previously according to which such hidden variables must exist. Such a contradiction would have allowed Schrödinger to reach a further conclusion which is quite striking, and which we shall discuss in Chap. 4.

References

1. Bell, J.S.: The theory of local beables. Epistemol. Lett. **9** (1976). Reprinted in [68].
2. Born, M. (eds.): The Born–Einstein Letters. Macmillan, London (1971)
3. Schilpp, P.A.: Albert Einstein: Philosopher-Scientist. Harper and Row, New York (1949)
4. von Neumann, J.: Mathematische Grundlagen der Quantenmechanik. Springer, Berlin (1932) English translation: Beyer, R.: Mathematical Foundations of Quantum Mechanics, pp. 305–324 (trans: Princeton University Press, Princeton) (1955)
5. Gleason, A.M.: J. Math. Mech. **6**, 885 (1957)
6. Kochen, S., Specker, E.P.: The problem of hidden variables in quantum mechanics. J. Math. Mech. **17**, 59 (1967)
7. Bell, J.S.: On the Einstein Podolsky Rosen paradox. Physics **1**, 195–200 (1964). Reprinted in [57], p. 14 and in [69], p. 403
8. Conway, J., Kochen, S.: The free will theorem. Found. Phys. **36**(10), 1441–1473 (2006). Physics archives arXiv:quantph/0604079
9. Conway, J., Kochen, S.: The strong free will theorem. Notices AMS **56**(2), 226–232 (2009)
10. Hemmick, D.: Doctoral dissertation, Rutgers University, Department of Physics (1996). Research supervised by Goldstein, S., Lebowitz, J. See arXiv:quant-ph/0412011v1
11. Bell, J.S.: Atomic cascade photons and quantum mechanical nonlocality. Comm. At. Mol. Phys. **9**, 121–126 (1980). Reprinted in [68], p. 782
12. Bell, J.S.: Bertlemann's socks and the nature of reality. J. de Physique Colloque C2. Tome 42 (suppl. au numero 3) C2 41–C2 61 (1981). Reprinted in [57], p. 139
13. Dürr, D., Goldstein, S., Zanghí, N.: Quantum equilibrium and the role of operators as observables in quantum theory. J. Stat. Phys. **116**, 959–1055 (2004)
14. Maudlin, T.: Quantum Non-locality and Relativity: Metaphysical Intimations of Modern Physics, 3rd edn. Wiley-Blackwell, Oxford (2011)
15. Maudlin, T.: Space-time in the quantum world. In: Fine, A., Goldstein, S., Cushing, J. (eds.) Bohmian Mechanics and Quantum Theory: An Appraisal, pp. 285–307. Kluwer, Dordrecht (1996)
16. Norsen, T.: Found. Phys. **37**, 311–340 (2007)
17. Norsen, T.: Bell locality and the nonlocal character of nature. Found. Phys. Lett. **19**(7), 633–655 (2006)
18. Norsen, T.: Local causality and completeness: Bell vs. Jarrett. Found. Phys. **39**(3), 273 (2009)
19. Wiseman, H.M.: From Einstein's theorem to Bell's theorem: A history of quantum nonlocality. Contemp. Phys. **47**, 79–88 (2006)
20. Jarrett, J.: On the physical significance of the locality conditions in the bell argument. Nous **18**, 569–589 (1984)
21. Evans, P., Price, H., Wharton, K.B.: New slant on the EPR-Bell experiment. The Physics archives arXiv:1001.5057v3 [quant-ph]

22. Stapp, H.: Quantum Nonlocality and the Description of Nature. In: Cushing J., McMullin, E. (eds.) Philosophical Consequences of Quantum Theory: Reflections on Bell's Theorem, pp.154–174. University of Notre Dame Press, Notre Dame

23. Redhead, M.: Incompletness, Nonlocality and Realism: A Prolegomenon to the Philosophy of Quantum Mechanics. Clarendon Press, Oxford (1987)

24. Davies, P.C.W., Brown, J.R.: John bell (interview) p. 45. In: The Ghost in the Atom. Cambridge University Press, Cambridge (1986)

25. Schrödinger, E.: Naturwissenschaften **23**, 807–812, 823–828, 844–849 (1935). English translation: Schrödinger, E.: The present situation in quantum mechanics. Proc. Cambridge Phil. Soc. **124**, 323–338 (1980) (trans: Drimmer, J.) and can also be found in [69]

26. Schrödinger, E.: Discussion of probability relations between separated systems. Proc. Cambridge Phil. Soc. **31**, 555 (1935)

27. Schrödinger, E.: Probability relations between separated systems. Proc. Cambridge Phil. Soc. **32**, 446 (1936)

28. Bassi, A., Ghirardi, G.C.: The Conway-Kochen argument and relativistic GRW Models. Found. Phys. **37**, 169 (2007). Physics archives: arXiv:quantph/0610209.

29. Tumulka, R.: Comment on 'The Free will Theorem'. Found. Phys. **37**, 186–197 (2007). Physics Archives arXiv:quant-ph/0611283.

30. Goldstein, S., Tausk, D., Tumulka, R., Zanghí, N.: What does the free will theorem actually prove? Notices of the AMS **57**(11), 1451–1453, arXiv:0905.4641v1 [quant-ph] (2010)

31. de Broglie, L.: Sur la possibilite de relier les phenomenes d'interference et de diffraction a la theorie des quanta de lumiere. Compt. Rend. **183**, 447–448 (1926)

32. de Broglie, L.: Sur la possibilite de mettre en accord la theorie electomagnetique avec la nouvelle. Compt. Rend. **185**, 380–382 (1927)

33. de Broglie, L.: Rapport au V'ieme Congres de Physique Solvay Gauthier–Villars, Paris (1930)

34. de Broglie, L.: Physicien et Penseur, p. 465, Gauthier-Villars, Paris (1953).

35. Bohm, D.: A suggested interpretation of the quantum theory in terms of "Hidden" variables. Phys. Rev. **85**, 166, 180 (1952)

36. Bell, J.S.: On the impossible pilot wave. Found. Phys. **12**, 989–999 (1982). Reprinted in [57], p. 159

37. Bell, J.S.: Six possible worlds of quantum mechanics. Proceedings of the Nobel Symposium 65: Possible Worlds in Arts and Sciences. Stockholm, 11–15 August 1986. Reprinted in [57], p. 181

38. Holland, P.R.: The Quantum Theory of Motion: An account of the de Broglie–Bohm Causal Interpretation of Quantum Mechanics. Cambridge University Press, Cambridge (1993). Reprinted in paperback 1995

39. Dürr, D., Teufel, S.: Bohmian Mechanics: The Physics and Mathematics of Quantum Theory. Springer, Berlin (2009). Reprinted 2010 in softcover

40. Cushing, J.T., Fine, A., Goldstein, S. (eds.): Bohmian Mechanics and Quantum Theory: An Appraisal. Springer, Berlin (1996). Reprinted in softcover 2010

41. Bell, J.S.: De Broglie-Bohm, delayed choice double slit experiment, and density matrix. Int. J. Quant. Chem. Quantum Chemistry Symposium. **14**, 155–159 (1980). Reprinted in [57], p. 111

42. Ghirardi, G.C, Rimini, A., Weber, T.: Unified dynamics for microscopic and macroscopic systems. Phys. Rev. D **34**, 470–491 (1986)

43. Bell, J.S.: Against 'Measurement'. Phys. World **3**, 33 (1990). This is reprinted in [68], p. 902

44. Bell, J.S.: Are there quantum jumps? Schrödinger, Centenary of a Polymath. Cambridge University Press, Cambridge (1987). Reprinted in [68], p. 866

45. Albert, David: Quantum Mechanics and Experience. Harvard University Press, Cambridge (1992)

46. Ghirardi, G.C.: Sneaking a Look at God's Cards Princeton. University Press, Princeton (2005)

47. Daumer, M., Dürr, D., Goldstein, S., Zanghí, N.: Naive realism about operators. In: Costantini, D., Gallavotti, M.C. (eds.) Erkenntnis, 1996, Special Issue in Honor of Prof. R. Jeffrey,

Proceedings of the International Conference "Probability, Dynamics and Causality", Luino, Italy 15–17 June 1995.

48. Greenberger, D.M., Horne, M.A., Zeilinger, A. Going beyond Bell's theorem. in [70], p. 69
49. Greenberger, D.M., Horne, M.A., Shimony, A., Zeilinger, A.: Bell's theorem without inequalities. Am. J. Phys. **58**, 1131–1143 (1990)
50. Mermin, N.D.: What's wrong with these elements of reality? Phys. Today **43**, 9 (1990)
51. Brown, H.R., Svetlichny, G.: Nonlocality and Gleason's lemma. Part I. Deterministic theories. Found. Phys. **20**, 1379 (1990)
52. Heywood, P., Redhead, M.L.G.: Nonlocality and the Kochen-Specker paradox. Found. Phys. **13**, 481 (1983)
53. Aravind, P.K.: Bell's theorem without inequalities and only two distant observers. Found. Phys. Lett. **15**, 399–405 (2002)
54. Cabello, A.: Bell's theorem without inequalities and without probabilities for two observers. Phys. Rev. Lett. **86**, 1911–1914 (2001)
55. Bell, J.S.: On the problem of hidden variables in quantum mechanics. Rev. Mod. Phys. **38**, 447–452 (1966). Reprinted in [57], p. 1 and [69]
56. Bell, J.S.: Introduction to the hidden-variable question. In: Foundations of Quantum Mechanics. Proceedings of the International School of Physics 'Enrico Fermi' course IL, New York, Academic, pp. 171–181 (1971). This article is reprinted in [57], p. 29
57. Bell, J.S.: Speakable and Unspeakable in Quantum Mechanics. Cambridge University Press, Cambridge (1987)
58. Bohm, D., Hiley, B.J.: The Undivided Universe: An Ontological Interpretation of Quantum Theory. Routledge, New York (1993)
59. Belinfante, F.J.: A Survey of Hidden-Variables Theories. Pergamon Press, New York (1973)
60. Hughes, R.I.G.: The Structure and Interpretation of Quantum Mechanics. Harvard University Press, Cambridge (1989)
61. Jammer, M.: The Philosophy of Quantum Mechanics. Wiley, New York (1974)
62. Albertson, J.: von Neumann's hidden - parameter proof. Am. J. Phys. **29**, 478 (1961)
63. Jauch, J.M., Piron, C.: Can hidden variables be excluded in quantum mechanics? Helv. Phys. Acta **36**, 827 (1963)
64. Einstein A., Podolsky B., Rosen N.: Can quantum mechanical description of physical reality be considered complete? Phys. Rev. **47**, 777 (1935). Reprinted in [69], p. 138
65. Schiff, L.: Quantum Mechanics. McGraw-Hill, New York (1955)
66. Shimony, A.: Search for a Naturalistic World View, vol. 2, p. 89. Cambridge University Press, Cambridge (1993)
67. Gilder, L.: The Age of Entanglement: When Quantum Physics Was Reborn. Alfred A Knopf, New York (2008)
68. Bell, M., Gottfried, K., Veltman, M., (eds.): Quantum Mechanics, High Energy Physics and Accelerators: Selected Papers of John S. Bell (with commentary). World Scientific Publishing Company, Singapore (1995), p. 744
69. Wheeler, J.A., Zurek, W.H. (eds.): Quantum Theory and Measurement. Princeton University Press, Princeton (1983), p. 397
70. Kafatos, M. (ed.): Bell's Theorem: Quantum Theory and Conceptions of the Universe. Kluwer Academic, Dordrecht (1989)

Chapter 2
Contextuality

We have seen in the previous chapter that the analysis of von Neumann has little impact on the question of whether a viable hidden variables theory may be constructed. However, further mathematical results were developed by Gleason [1] in 1957 and by Simon Kochen and Specker in 1967 [2], which were claimed by some[1] to imply the impossibility of hidden variables. In the words of Kochen and Specker [2, p. 73]: "If a physicist X believes in hidden variables... ... the prediction of X contradicts the prediction of quantum mechanics". The Gleason, and Kochen and Specker arguments are in fact, *stronger* than von Neumann's in that they assume linearity only for *commuting* observables. Despite this, a close analysis reveals that the impossibility proofs of Gleason and of Kochen and Specker share with von Neumann's proof the neglect of the possibility of a hidden variables feature called *contextuality* [5, 6]. We will find that this shortcoming makes these theorems inadequate as proofs of the impossibility of hidden variables.

We begin this chapter with the presentation of Gleason's theorem and the theorem of Kochen and Specker. This will be followed by a discussion of contextuality and its relevance to these analyses. We will make clear in this discussion why the theorems in question fail as impossibility proofs. As far as the question of what conclusions *do* follow, we show that these theorems' implications can be expressed in a simple and concise fashion, which we refer to as "spectral-incompatibility". We conclude the chapter with the discussion of an experimental procedure first discussed by Albert[2] which provides further insight into contextuality.

---

[1] See [2–4].

[2] See Albert [7].

D. L. Hemmick and A. M. Shakur, *Bell's Theorem and Quantum Realism*,
SpringerBriefs in Physics, DOI: 10.1007/978-3-642-23468-2_2,
© The Author(s) 2012

2.1 Gleason's Theorem

Von Neumann's theorem addressed the question of the form taken by a function $E(O)$ of the observables. Gleason's theorem essentially addresses the same question,[3] the most significant difference being that the linearity assumption is relaxed to the extent that it is demanded that E be linear on only *commuting* sets of observables. Besides this, Gleason's theorem involves a function E on only the projection operators of the system, rather than on all observables. Finally, Gleason's theorem contains the assumption that the system's Hilbert space is at least three dimensional. As for the conclusion of the theorem, this is identical to von Neumann's: $E(P)$ takes the form $E(P) = \text{Tr}(UP)$ where U is a positive operator and $\text{Tr}(U) = 1$.

Let us make the requirement of linearity on the commuting observables somewhat more explicit. First we note that any set of projection $\{P_1, P_2, \ldots\}$ onto mutually orthogonal subspaces $\{\mathcal{H}_1, \mathcal{H}_2, \ldots\}$ will form a commuting set. Furthermore, if P projects onto the direct sum $\mathcal{H}_1 \oplus \mathcal{H}_2 \oplus \ldots$ of these subspaces, then $\{P, P_1, P_2, \ldots\}$ will also form a commuting set. It is in the case of this latter type of set that the linearity requirement comes into play, since these observables obey the relationship

$$P = P_1 + P_2 + \cdots. \tag{2.1}$$

The condition on the function E is then

$$E(P) = E(P_1) + E(P_2) + \cdots. \tag{2.2}$$

The formal statement of Gleason's theorem is expressed as follows. For any quantum system whose Hilbert space is at least three dimensional, any expectation function $E(P)$ obeying the conditions (2.2), $0 \leq E(P) \leq 1$, and $E(\mathbf{1}) = 1$ must take the form

$$E(P) = \text{Tr}(UP), \tag{2.3}$$

where $\text{Tr}(U) = 1$ and U is a positive operator. We do not present the proof of this result[4] here. In the next section, we present an outline the proof of Kochen and Specker's theorem. The same impossibility result derivable from Gleason's work also follows from this theorem.

---

[3]   The original form presented by A.M. Gleason referred to a probability measure on the subspaces of a Hilbert space, but the equivalence of such a construction with a value map on the projection operators is simple and immediate. This may be seen by considering that there is a one-to-one correspondence between the subspaces and projections of a Hilbert space and that the values taken by the projections are 1 and 0, so that a function mapping projections to their eigenvalues is a special case of a probability measure on these operators.

[4]   See Bell [5]. Bell proves that any function $E(P)$ satisfying the conditions of Gleason's theorem cannot map the projection operators to their eigenvalues.

It is straightforward to demonstrate that the function $E(P)$ considered within Gleason's theorem cannot be a value map function on these observables. To demonstrate this, one may argue in the same fashion as was done by von Neumann, since the form developed here for $E(P)$ is the same as that concluded by the latter. (See Sect. 1.4.3). We recall that if $E(O)$ is to represent a dispersion free state specified by ψ and λ, it *must* take the form of such a value map, and $E(O)$ evidently cannot be the expectation function for such a state. It is on this basis that the impossibility of hidden variables has been claimed to follow from Gleason's theorem.

2.2 Kochen and Specker's Theorem

As with Gleason's theorem, the essential assumption of Kochen and Specker's theorem is that the expectation function $E(O)$ must be linear on commuting sets of observables. It differs from the former only in the set of observables considered. Gleason's theorem was addressed to the projection operators on a Hilbert space of arbitrary dimension N. Kochen and Specker consider the squares $\{s_{\theta,\phi}^2\}$ of the spin components of a spin 1 particle. One may note that these observables are formally identical to projection operators on a three-dimensional Hilbert space. Thus, the Kochen and Specker observables are a subset[5] of the "$N = 3$" case of the Gleason observables. Among the observables $\{s_{\theta,\phi}^2\}$ any subset $\{s_x^2, s_y^2, s_x^2\}$ corresponding to mutually orthogonal component directions x, y, z will be a commuting set. Each such set obeys the relationship

$$s_x^2 + s_y^2 + s_z^2 = 2. \tag{2.4}$$

For every such subset, Kochen and Specker require that $E(s_{\theta,\phi}^2)$ must obey

$$E(s_x^2) + E(s_y^2) + E(s_z^2) = 2. \tag{2.5}$$

Kochen and Specker theorem states that there exists no function $E(s_{\theta,\phi}^2)$ on the squares of the spin components of a spin 1 particle which maps each observable to either 0 or 1 and which satisfies (2.5).

We now make some comments regarding the nature of this theorem's proof. The problem becomes somewhat simpler to discuss when formulated in terms of a geometric model. Imagine a sphere of unit radius surrounding the origin in $\mathbb{R}^3$. It is easy to see that each point on this sphere's surface corresponds to a direction in space, which implies that each point is associated with one observable of the set $\{s_{\theta,\phi}^2\}$. With this, E may be regarded as a function on the surface of the unit sphere. Since the eigenvalues of each of these spin observables are 0 and 1, it must be that $E(O)$ must take on these values, if it is to assume the form of a value map function. Satisfaction of (2.5) requires that for each set of mutually orthogonal directions, E must assign to

---

[5] The set of projections on a three-dimensional space is actually a larger class of observables.

one of them 0 and to the other directions 1. To gain some understanding[6] of why such an assignment of values must fail, we proceed as follows. To label the points on the sphere, we imagine that each point on the sphere to which the number 0 is assigned is painted red, and each point assigned 1 is painted blue. We label each direction as by the unit vector $\hat{n}$. Since one direction of every mutually perpendicular set is assigned red, then in total we require that one-third of the sphere is painted red. If we consider the components of the spin in *opposite* directions θ, ϕ, and $180° - \theta$, $180 + \phi$, these values are always opposite, i.e., if s_x takes the value $+1$, then s_{-x} takes the value -1. This implies that the values of $s^2_{\theta,\phi}$ and $s^2_{180°-\theta,180+\phi}$ will be *equal*. Therefore, we must have that points on the sphere lying directly opposite each other, i.e., the "antipodes", must receive the same assignment from E. Suppose that one direction and its antipode are painted red. These points form the two poles of a great circle, and all points along this circle must then be painted blue, since all such points represent directions orthogonal to the directions of our two 'red' points. For every such pair of red points on the sphere, there must be many more blue points introduced, and we will find this makes it impossible to make one-third of the sphere red, as would be necessary to satisfy (2.5).

Suppose we paint the entire first octant of the sphere red. In terms of the coordinates used by geographers, this is similar to the region in the Northern hemisphere between 0 and 90° longitude. If the point at the north pole is painted red then the great circle at the equator must be blue. Suppose that the 0 and 90° meridians are also painted red. Then the octant which is the antipode of the first octant must also be painted red. This octant would be within the 'southern hemisphere' between 180 and 270° longitude. If we now apply the condition that for every point painted red, all points lying on the great circle defined with the point at its pole, we find that all remaining points of the sphere must then be colored blue, thereby preventing the addition of more red points. This assignment implies that more than two-thirds of the sphere is blue, therefore some sets of mutually perpendicular directions are all colored blue. An example of such a set of directions is provided by the points on the sphere's surface lying in the mutually orthogonal directions represented by $(0.5, 0.5, -0.7071)$, $(-0.1464, 0.8535, 0.5)$, $(0.8535, -0.1464, 0.5)$. Each of these points lies in a quadrant to which we have assigned the color blue by the above scheme. This assignment of values to the spatial directions must therefore fail to meet the criteria demanded of the function $E(s^2_{\theta,\phi})$ in Kochen and Specker's premises: that $E(s^2_{\theta,\phi})$ satisfies (2.5) and maps the observables to their values.

In their proof, Kochen and Specker show that for a discrete set of 117 different directions in space, it is impossible to give appropriate value assignments to the corresponding spin observables.[7] Kochen and Specker then assert that hidden variables cannot agree with the predictions of quantum mechanics. Their conclusion is that if some physicist 'X', mistakenly decides to accept the validity of hidden variables then

---

[6] We follow here the argument given in Belinfante [8, p. 38]

[7] Since their presentation, the proof has been simplified by Peres in 1991 [9] whose proof is based on examination of 33 such $\hat{n}$ vectors. We also note that a proof presented by Bell [5] may be shown to lead easily to a proof of Kochen and Specker. See Mermin in this connection [10].

"the prediction of X (for some measurements) contradicts the prediction of quantum mechanics" [2, p 73]. The authors cite a particular system on which one can perform an experiment they claim reveals the failure of the hidden variables prediction. We will demonstrate in the next section of this work that what follows from Kochen and Specker's theorem is only that a *non-contextual* hidden variables theory will conflict with quantum mechanics, so that the general possibility of hidden variables has not been disproved. Furthermore, we show that if the type of experiment envisioned by these authors is considered in more detail, it does not indicate where hidden variables must fail, but instead serves as an illustration indicating that the requirement of contextuality is a quite natural one.

More recent proofs have been offered involving smaller numbers of spin component observables rather than the 117 utilized by the original Kochen and Specker proof. In particular, Asher Peres has shown[8] that value assignments to a particular set of 33 spin observables cannot be made such that quantum mechanical prescriptions are met.

2.3 Contextuality and Gleason's, and Kochen and Specker's Impossibility Proofs

We have seen that the theorems of Gleason, and Kochen and Specker each demonstrate the impossibility of value maps on some sets of observables such that the constraining relationships on each commuting set are obeyed. One might be at first inclined to conclude with Kochen and Specker that such results imply the impossibility of a hidden variables theory. However, if we consider that there exists a successful theory of hidden variables, namely Bohmian mechanics [12] (see Sect. 1.1.2), we see that such a conclusion is in error. Moreover, an explicit analysis of Gleason's theorem has been carried out by Bell [5, 6] and its inadequacy as an impossibility proof was shown. Bell's argument may easily be adapted[9] to provide a similar demonstration regarding Kochen and Specker's theorem. The key concept underlying Bell's argument is that of *contextuality*, and we now present a discussion of this notion.

Essentially, contextuality refers to the dependence of measurement results on the detailed experimental arrangement being employed. In discussing this notion, we will find that an inspection of the quantum formalism suggests that contextuality is a natural feature to expect in a theory explaining the quantum phenomena. Furthermore we shall find that the concept is in accord with Niels Bohr's remarks regarding the fundamental principles of quantum mechanics. According to Bohr [13] "*a closer examination reveals that the procedure of measurement has an essential influence on the conditions on which the very definition of the physical quantities in question*

---

[8] See [11].

[9] As we have mentioned, since the observables are formally equivalent to projections on a three-dimensional Hilbert space, this theorem is actually a special case of Gleason's. Therefore, Bell's argument essentially addresses the Kochen and Specker theorem as well as Gleason's.

rests." In addition, he stresses [14, p. 210] *"the impossibility of any sharp distinction between the behavior of atomic objects and the interaction with the measuring instruments which serve to define the conditions under which the phenomena appear.*" The concept of contextuality represents a concrete manifestation of the quantum theoretical aspect to which Bohr refers. We will first explain the concept itself in detail, and then focus on its relevance to the theorems of Gleason, and Kochen and Specker.

We begin by recalling a particular feature of the quantum formalism. In the presentation of this formalism given in chapter one, we discussed the representation of the system's state, the rules for the state's time evolution, and the rules governing the measurement of an observable. The measurement rules are quite crucial, since it is only through measurement that the physical significance of the abstract quantum state (given by ψ) is made manifest. Among these rules, one finds that any commuting set of observables may be measured simultaneously. With a little consideration, one is led to observe that the possibility exists for *a variety of different experimental procedures* to measure a single observable. Consider for example, an observable O which is a member of the commuting set $\{O, A_1, A_2, \ldots\}$. We label this set as C. A simultaneous measurement of the set C certainly gives among its results a value for O and thus may be regarded as providing a measurement of O. It is possible that O may be a member of another commuting set of observables $C' = \{O, B_1, B_2, \ldots\}$, so that a simultaneous measurement of C' also provides a measurement of O. Let us suppose further that the members of set $\{A_i\}$ fail to commute with those of $\{B_i\}$. It is then clear that experiments measuring C and C' are quite different, and hence must be distinct. A concrete difference appears for example, in the effects of such experiments on the system wave function. The measurement rules tell us that the wave function subsequent to an ideal measurement of a commuting set is prescribed by the Eq. 1.12, according to which the post-measurement wave function is calculated from the pre-measurement wave function by taking the projection of the latter into the joint-eigenspace of that set. Since the members of C and C' fail to commute, the joint-eigenspaces of the two are necessarily different, and the system wave function will not generally be affected in the same way by the two experimental procedures. Apparently the concept of 'the measurement of an observable' is *ambiguous*, since there can be distinct experimental procedures for the measurement of a single observable.

There are, in fact, more subtle distinctions between different procedures for measuring the same observable, and these also may be important. To introduce the experimental procedure of measurement into our formal notation, we shall write $\mathcal{E}(O)$, $\mathcal{E}'(O)$, etc., to represent experimental procedures used to measure the observable O. From what we have seen here, it is quite natural to expect that a hidden variables theory should allow for the possibility that *different experimental procedures*, e.g., $\mathcal{E}(O)$ and $\mathcal{E}'(O)$, for the measurement of an observable might yield *different results* on an individual system. This is contextuality.

Examples of observables for which there exist incompatible measurement procedures are found among the observables addressed in each of the theorems of Gleason, and Kochen and Specker. Among observables addressed by Gleason are the one-dimensional projection operators $\{P_\phi\}$ on an N-dimensional Hilbert space $\mathcal{H}_N$.

Consider a one-dimensional projection P_ϕ where ϕ belongs to two sets of orthonormal vectors given by $\{\phi, \psi_1, \psi_2, \ldots\}$ and $\{\phi, \chi_1, \chi_2, \ldots\}$. Note that the sets $\{\psi_1, \psi_2, \ldots\}$ and $\{\chi_1, \chi_2, \ldots\}$ are constrained only in that they span $\mathcal{H}_\phi^\perp$ (the orthogonal complement of the one-dimensional space spanned by ϕ). Given this, there exist examples of such sets for which some members of $\{\psi_1, \psi_2, \ldots\}$ are distinct from and not orthogonal to the vectors in $\{\chi_1, \chi_2, \ldots\}$. Since any distinct vectors that are not orthogonal correspond to projections which fail to commute, the experimental procedures $\mathcal{E}(P_\phi)$ measuring $\{P_\phi, P_{\psi_1}, P_{\psi_2}, \ldots\}$ and $\mathcal{E}'(P_\phi)$ measuring $\{P_\phi, P_{\chi_1}, P_{\chi_2}, \ldots\}$ are incompatible. The argument just given applies also to the Kochen and Specker observables (the squares of the spin components of a spin 1 particle), since these are formally identical to projections on a three-dimensional Hilbert space. To be explicit, the observable s_x^2 is a member of the commuting sets $\{s_x^2, s_y^2, s_z^2\}$ and $\{s_x^2, s_{y'}^2, s_{z'}^2\}$ where the y' and z' are oblique relative to the y and z axes. In this case, $s_{y'}^2, s_{z'}^2$ do not commute with s_y^2, s_z^2. Thus, the experimental procedures to measure these sets are incompatible.

While it is true that the arguments against hidden variables derived from these theorems are superior to von Neumann's, since they require agreement only with operator relationships among commuting sets, these arguments nevertheless possess the following shortcoming. Clearly, the mathematical functions considered in each case, $E(P)$ and $E(s_{\theta, \phi}^2)$ do *not* allow for the possibility that the results of measuring each observable using different and possibly *incompatible* procedures may lead to different results. What the theorems demonstrate is that no hidden variables formulation *based on assignment of a unique value to each observable* can possibly agree with quantum mechanics. But this is a result we might well have expected from the fact that the quantum formalism allows the possibility of incompatible experimental procedures for the measurement of an observable. For this reason, neither of the theorems here considered—Gleason's theorem, and Kochen and Specker's theorem—imply the impossibility of hidden variables, since they fail to account for such a fundamental feature of the quantum formalism's rules of measurement.

2.3.1 Procedure to Measure the Kochen and Specker Observables

In a discussion of the implications of their theorem, Kochen and Specker mention a system for which well-known techniques of atomic spectroscopy may be used to measure the relevant spin observables. Although these authors mention this experiment to support their case against hidden variables, the examination of such an experiment actually reinforces the assertion that one should allow for contextuality—the very concept that refutes their argument against hidden variables.

Kochen and Specker note[10] that for an atom of *orthohelium*[11] which is subjected to an electric field of a certain configuration, the first-order effects of this field on the electrons may be accounted for by adding a term of the form $aS_x^2 + bS_y^2 + cS_z^2$ to the electronic Hamiltonian. Here a, b, c are distinct constants, and S_x^2, S_y^2, S_z^2 are the squares of the components of the total spin of the two electrons with respect to the Cartesian axes x, y, z. The Cartesian axes are defined by the orientation of the applied external field. For such a system, an experiment measuring the energy of the electrons also measures the squares of the three spin components. To see this, note that the value of the perturbation energy will be $(a + b)$, $(a + c)$, or $(b + c)$ if the joint values of the set S_x^2, S_y^2, S_z^2 equal respectively $\{1, 1, 0\}$, $\{1, 0, 1\}$, or $\{0, 1, 1\}$.

To understand why the external electric field affects the orthohelium electrons in this way, consider the ground state of orthohelium.[12] The wave function of this state is given by a spatial part $\phi(r_1, r_2)$, depending only on r_1, r_2 (the radial coordinates of the electrons), multiplied by the spin part, which is a linear combination of the eigenvectors $\psi_{+1}, \psi_0, \psi_{-1}$ of S_z, corresponding to $S_z = +1, 0, -1$, respectively. Thus, the ground state may be represented by any vector in the three-dimensional Hilbert space spanned by the vectors $\phi(r_1, r_2)\psi_{+1}$, $\phi(r_1, r_2)\psi_0$, $\phi(r_1, r_2)\psi_{-1}$. The external electric field will have the effect of "lifting the degeneracy" of the state, i.e., the new Hamiltonian will not be degenerate in this space, but its eigenvalues will correspond to three unique orthogonal vectors. Suppose that we consider a particular set of Cartesian axes x, y, z. We apply an electric field which is of orthorhombic symmetry[13] with respect to these axes. It can be shown[14] that the eigenvectors of the Hamiltonian

---

[10] One can derive the analogous first-order perturbation term arising for a charged particle of orbital angular momentum $L = 1$ in such an electric field using the fact that the joint-eigenstates of L_x^2, L_y^2, L_y^2 are the eigenstates of the potential energy due to the field. This latter result is shown in Kittel [15, p. 427].

[11] Orthohelium and parahelium are two species of helium which are distinguished by the total spin S of the two electrons: for the former we have $S = 1$, and for the latter $S = 0$. There is a rule of atomic spectroscopy which prohibits atomic transitions for which $\Delta S = 1$, so that no transitions from one form to the other can occur spontaneously.

[12] Using spectroscopic notation, this state would be written as the '2^3 S' state of orthohelium. The '2' refers to the fact that the principal quantum number n of the state equals 2, 'S' denotes that the total orbital angular momentum is zero, and the '3' superscript means that it is a spin triplet state. Orthohelium has no state of principal quantum number $n = 1$, since the Pauli exclusion principle forbids the '1^3 S' state.

[13] Orthorhombic symmetry is defined by the criterion that rotation about either the x or y axis by $180°$ would bring such a field back to itself.

[14] A straightforward way to see this is by analogy with a charged particle of orbital angular momentum $L = 1$. The effects of an electric or magnetic field on a charged particle of spin 1 are analogous to the effects of the same field on a charged particle of orbital angular momentum 1. To calculate the first-order effects of an electric field of orthorhombic symmetry for such a particle, one can examine the spatial dependence of the $L_z = 1, 0, -1$ states $\psi_{-1}, \psi_0, \psi_{+1}$, together with the spatial dependence of the perturbation potential $V(\mathbf{r})$, to show that the states $1/\sqrt(\psi_1 - \psi_{-1})$, $1/\sqrt{2}(\psi_1 + \psi_{-1})$, and ψ_0 are the eigenstates of such a perturbation. A convenient choice of V for this purpose is $V = Ax^2 + By^2 + Cz^2$. See Kittel in [15, p. 427].

due to this field are $v_1 = 1/\sqrt{2}((\psi_1 - \psi_{-1})$, $v_2 = 1/\sqrt{2}(\psi_1 + \psi_{-1})$ and $v_3 = \psi_0$. We drop the factor $\phi(r_1, r_2)$ for convenience of expression. These vectors are *also* the joint-eigenvectors of the observables S_x^2, S_y^2, S_z^2, as we can easily show. When expressed as a matrix in the $\{\psi_{+1}, \psi_0, \psi_{-1}\}$ basis, the vectors v_1, v_2, v_3 take the form

$$v_1 = \begin{pmatrix} \frac{1}{\sqrt{2}} \\ 0 \\ -\frac{1}{\sqrt{2}} \end{pmatrix} \tag{2.6}$$

$$v_2 = \begin{pmatrix} \frac{1}{\sqrt{2}} \\ 0 \\ \frac{1}{\sqrt{2}} \end{pmatrix} \tag{2.7}$$

$$v_3 = \begin{pmatrix} 0 \\ 1 \\ 0 \end{pmatrix}. \tag{2.8}$$

If we then express S_x^2, S_y^2, S_z^2 as matrices in terms of the same basis, then by elementary matrix multiplication, one can show that v_1 corresponds to the joint-eigenvalue $\mu = \{0, 1, 1\}$, v_2 corresponds to $\mu = \{1, 0, 1\}$, and v_3 corresponds to $\mu = \{1, 1, 0\}$. Thus, the eigenvectors v_1, v_2, v_3 of the Hamiltonian term H' which arises from a perturbing electric field (defined with respect to x, y, z) are also the joint-eigenvectors of the set $\{S_x^2, S_y^2, S_z^2\}$. This implies that we can represent H' by the expression $aS_x^2 + bS_y^2 + cS_z^2$, where H''s eigenvalues are $\{(b + c), (a + c), (a + b)\}$.

All of this leads to the following conclusion regarding the measurement of the spin of the orthohelium ground state electrons. Let the system be subjected to an electric field with orthorhombic symmetry with respect to a given set of Cartesian axes x, y, z. Under these circumstances, the measurement of the total Hamiltonian will yield a result equal (to first-order approximation) to the (unperturbed) ground state energy plus one of the perturbation corrections $\{(b + c), (a + c), (a + b)\}$. If the measured value of the perturbation energy is $(a + b)$, $(a + c)$, or $(b + c)$ then the joint values of the set $\{S_x^2, S_y^2, S_z^2\}$ are given respectively by $\{1, 1, 0\}$, $\{1, 0, 1\}$, or $\{0, 1, 1\}$.

It is quite apparent from this example that it would be unreasonable to require that a hidden variables theory must assign a single value to S_x^2, independent of the experimental procedure.

2.4 Contextuality Theorems and Spectral Incompatibility

We saw in our discussion of von Neumann's theorem that its implications toward hidden variables amounted to the assertion that there can be no mathematical function $E(O)$ that is linear on the observables and which maps them to their eigenvalues. This was neither a surprising, nor particularly enlightening result, since it follows also from a casual observation of some example of linearly related non-commuting observables, as we saw in examining the observables $1/\sqrt{2}(\sigma_x + \sigma_y), \sigma_x, \sigma_y$ of a spin $\frac{1}{2}$ particle. The theorems of Gleason, and Kochen and Specker, imply a somewhat less obvious type of impossibility: there exists no function $E(O)$ mapping the observables to their eigenvalues, which obeys all relationships constraining *commuting* observables. What we develop here is a somewhat simpler expression of the implication of these theorems. We will find that they imply the *spectral-incompatibility* of the value map function: there exists no mathematical function that assigns to each commuting set of observables a joint-eigenvalue of that set.

We begin by recalling the notions of joint eigenvectors and joint eigenvalues of a commuting set of observables $(O^1, O^2, \ldots)$. For a commuting set, the eigenvalue Eq. 1.5 $O|\psi\rangle = \mu|\psi\rangle$ is replaced by a set of relationships (1.7): $O^i\psi = \mu^i|\psi\rangle$ $i = 1, 2, \ldots$, one for each member of the commuting set. If a given $|\psi\rangle$ satisfies this relationship for *all* members of the set, it is referred to as a *joint-eigenvector*. The set of numbers $(\mu^1, \mu^2, \ldots)$ that allow the equations to be satisfied for this vector are collectively referred to as the *joint-eigenvalue* corresponding to this eigenvector, and the symbol $\boldsymbol{\mu} = (\mu_1, \mu_2, \ldots)$ is used to refer to this set. The set of all joint-eigenvalues $\{\boldsymbol{\mu}_a\}$ is given the name 'joint-eigenspectrum'.

In general, the members of any given commuting set of observables might not be independent, i.e., they may be constrained by mathematical relationships. We label the relationships for any given commuting set $\{O^1, O^2, \ldots\}$ as

$$f_1(O^1, O^2, \ldots) = 0$$
$$f_2(O^1, O^2, \ldots) = 0 \qquad\qquad (2.9)$$
$$\vdots$$

The Eqs. 2.1 in Gleason's theorem, and 2.4 in Kochen and Specker's theorem are just such relations. We now demonstrate the following two results. First, that every member of the joint-eigenspectrum must satisfy all relationships (2.9). Second, that any set of numbers $\xi_1, \xi_2, \ldots$ satisfying all of these relationships is a joint-eigenvalue.

To demonstrate the first of these, we suppose that $\boldsymbol{\mu} = (\mu^1, \mu^2, \ldots)$ is a joint-eigenvalue of the commuting set $\{O^1, O^2, \ldots\}$, with joint-eigenspace $\mathcal{H}$. We then consider the operation of $f_i(O^1, O^2, \ldots)$ on a vector $\psi \in \mathcal{H}$ where $f_i(O^1, O^2, \ldots) = 0$ is one of the relationships constraining the commuting set. We find

$$f_i(O^1, O^2, \ldots)\psi = f_i(\mu_1, \mu_2, \ldots)\psi = 0. \qquad\qquad (2.10)$$

The second equality implies that $f_i(\mu_1, \mu_2, \ldots) = 0$. Since $f_i(O^1, O^2, \ldots) = 0$ is an arbitrary member of the relationships (2.9) for the commuting set $\{O^1, O^2, \ldots\}$, it follows that every joint-eigenvalue $\boldsymbol{\mu}$ of the set must satisfy *all* such relationships.

We now discuss the demonstration of the second point. Suppose that the numbers $\{\xi_1, \xi_2, \ldots\}$ satisfy all of (2.9) for some commuting set $\{O^1, O^2, \ldots\}$. We consider the following relation:

$$\left(\left[\left(O^1 - \mu_1^1\right)^2 + \left(O^2 - \mu_1^2\right)^2 + \ldots\right]\left[\left(O^1 - \mu_2^1\right)^2 + \left(O^2 - \mu_2^2\right)^2 + \ldots\right]\ldots\right)\psi = 0.$$
(2.11)

Here, we operate on the vector ψ with a product whose factors each consist of a sum of various operators. The product is taken over all joint-eigenvalues $\{\mu_i\}$. We represent each joint-eigenvalue μ_i by a set $\left(\mu_i^1, \mu_i^2, \ldots\right)$. The validity of (2.11) is easily seen. Since the joint-eigenspaces of any commuting set are complete, the vector ψ must lie within such a space. Suppose that $\psi \in \mathcal{H}_i$, where $\mathcal{H}_i$ is the joint-eigenspace corresponding to μ_i. Then the ith factor in the product of operators in (2.11) must give zero when operating on ψ. Therefore, the entire product operating on ψ must also give zero. Since ψ is an arbitrary vector, it follows that

$$\left[\left(O^1 - \mu_1^1\right)^2 + \left(O^2 - \mu_1^2\right)^2 + \ldots\right]\left[\left(O^1 - \mu_2^1\right)^2 + \left(O^2 - \mu_2^2\right)^2 + \cdots\right]\ldots = 0.$$
(2.12)

Note that (2.12) is itself a constraining relationship on the commuting set, so that it must be satisfied by the numbers $(\xi_1, \xi_2, \ldots)$. This can only be true if these numbers form a joint-eigenvalue of $\{O^1, O^2, \ldots\}$, and this is the result we were to prove.

From this, we can discern a simple way to understand the implications of the theorems of Gleason, and Kochen and Specker toward the question of a value map. The requirement (2.2) Gleason's theorem places on the function $E(P)$ can be re-stated as the requirement that for each commuting set, $E(P)$ must satisfy all the relationships constraining its members. From the above argument, it follows that this assumption is equivalent to the constraint that $E(P)$ must assign to each commuting set a joint-eigenvalue. The same is also true of assumption that $E(s_{\theta,\phi}^2)$ satisfy Eq. 2.5.

Thus, both of these theorems can be regarded as proofs of the impossibility of a function mapping the observables to their values such that each commuting set is assigned a joint-eigenvalue. An appropriate name for such a proof would seem to be 'spectral-incompatibility theorem.'

2.5 Albert's Example and Contextuality

In thinking about any given physical phenomenon, it is natural to try to picture to oneself the properties of the system being studied. In using the quantum formalism to develop such a picture, one may tend to regard the 'observables' of this formalism, i.e., the Hermitian operators (see Sect. 1.3), as representative of these properties. However, the central role played by the *experimental procedure* $\mathcal{E}(O)$ in the measurement of any given observable O seems to suggest that such a view of the operators may be untenable. We describe an experiment originally discussed by David Albert[15] that indicates that this is indeed the case: the Hermitian operators cannot be regarded

---

[15]   See Albert in [7]. The experiment is also discussed by Ghirardi in [16].

as representative of the properties of the system.[16] Albert considers two laboratory procedures that may be used to measure the z-component of the spin of a spin $\frac{1}{2}$ particle. Although the two procedures are quite similar to one another, they cannot be regarded as identical when considered in light of the hidden variables theory known as Bohmian mechanics. This is a particularly striking instance of contextuality, and it indicates the inadequacy of the conception that the spin operator σ_z represents an intrinsic property of the particle. From Albert's example, one can clearly see that the outcome of the σ_z measurement depends not only on the parameters of the particle itself, but also on the complete experimental setup.

The Albert example is concerned with the measurement of spin[17] as performed using a Stern–Gerlach magnet. The schematic diagram given in Fig. 2.1. exhibits the configuration used in both of the measurement procedures to be described here. Note that we use a Cartesian coordinate system for which the x-axis lies along the horizontal direction with positive x directed toward the right, and the z-axis lies along the vertical direction with positive z directed upward. The y-axis (not shown) is perpendicular to the plane of the figure, and—since we use a right-handed coordinate system—positive y points into this plane. The long axis of the Stern–Gerlach magnet system is oriented along the x-axis, as shown. The upper and lower magnets of the apparatus are located in the directions of positive z and negative z. We define the Cartesian system further by requiring that x-axis (the line defined by $y = 0, z = 0$) passes through the center of the Stern–Gerlach magnet system.

In each experiment, the spin $\frac{1}{2}$ particle to be measured is incident on the apparatus along the positive x-axis. In the region of space the particle occupies before entering the Stern–Gerlach apparatus, its wave function is of the form

$$\psi_t(\mathbf{r}) = \varphi_t(\mathbf{r})(|\uparrow\rangle + |\downarrow\rangle), \tag{2.13}$$

where the vectors $|\uparrow\rangle$ and $|\downarrow\rangle$ are the eigenvectors of σ_z corresponding to eigenvalues $+\frac{1}{2}$ and $-\frac{1}{2}$, respectively. Here $\varphi_t(\mathbf{r})$ is a localized wave packet moving in the positive x direction toward the magnet.

We wish to consider two experiments that differ only in the orientation of the magnetic field inside the Stern–Gerlach apparatus. In experiment 1, the upper magnet has a strong magnetic north pole toward the region of particle passage, while the lower has a somewhat weaker magnetic south pole toward this region. In experiment 2, the magnets are such that the gradient of the field points in the *opposite* direction, i.e., the upper magnet has a strong magnetic *south* pole toward the region of passage, while the lower has a weak magnetic north pole towards it. After passing the Stern–Gerlach apparatus, the particle will be described by a wave function of one of the following forms:

---

[16] This idea has been propounded by Daumer et al. in [17]. See also Bell in [18].

[17] As is usual in discussions of Stern–Gerlach experiments, we consider only those effects relating to the interaction of the magnetic field with the magnetic moment of the particle. We consider the electric charge of the particle to be zero.

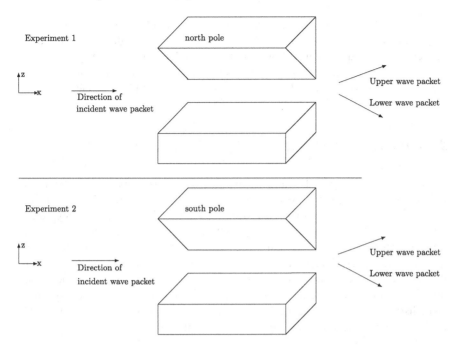

Fig. 2.1 Geometry of the Stern–Gerlach Experiments

$$\psi_t^1(\mathbf{r}) = \frac{1}{\sqrt{2}}(\phi_t^+(\mathbf{r})|\uparrow\rangle + \phi_t^-(\mathbf{r})|\downarrow\rangle)$$

$$\psi_t^2(\mathbf{r}) = \frac{1}{\sqrt{2}}(\phi_t^-(\mathbf{r})|\uparrow\rangle + \phi_t^+(\mathbf{r})|\downarrow\rangle).$$

(2.14)

Here $\psi_t^1(\mathbf{r})$ corresponds to experiment 1, and $\psi_t^2(\mathbf{r})$ corresponds to experiment 2. In both cases, the function $\phi_t^+(\mathbf{r})$ represents a localized wave packet moving obliquely upward and $\phi_t^-(\mathbf{r})$ represents a localized wave packet moving obliquely downward. To measure σ_z, one places detectors in the paths of these wave packets. Examination of the first equation of (2.14) shows that for experiment 1, if the particle is detected in the upper path, the result of our σ_z measurement is $+\frac{1}{2}$. If the particle is detected in the lower path, the result is $-\frac{1}{2}$. For experiment 2, the second equation of (2.14) leads to the conclusion that similar detections are associated with results opposite in sign to those of experiment 1. Thus, for experiment 2, detection in the upper path implies $\sigma_z = -\frac{1}{2}$, while detection in the lower implies $\sigma_z = +\frac{1}{2}$.

We make here a few remarks regarding the symmetry of the system. We constrain the form of the wave packet $\varphi_t(\mathbf{r})$ of (2.13), by demanding that it has no dependence on y, and that it exhibits reflection symmetry through the plane defined by $z = 0$, i.e., $\varphi_t(x, z) = \varphi_t(x, -z)$. Moreover, the vertical extent of this wave packet is to be the same size as the vertical spacing between the upper and lower magnets of the apparatus. As regards the wave packets $\phi_t^+(\mathbf{r})$ and $\phi_t^-(\mathbf{r})$ of five, if the magnetic field within the apparatus is such that $\partial B_z / \partial z$ is constant[18] then for both experiments, these packets move at equal angles above and below the horizontal (See Fig. 2.1). Thus, the particle is described both before and after it passes the Stern–Gerlach magnet, by a wave function which has reflection symmetry through the plane defined by $z = 0$.

2.5.1 Bohmian Mechanics and Albert's Example

We have mentioned that the hidden variables theory developed by David Bohm [12] gives an explanation of quantum phenomena which is empirically equivalent to that given by the quantum formalism. Bohmian mechanics allows us to regard any given system as a set of particles having well-defined (but distinctly non-Newtonian [20]) trajectories. Within Bohmian mechanics, it is the *configuration* of the system $\mathbf{q} = (q_1, q_2, q_3, ...)$ which plays the role of the hidden variables parameter λ. Thus, the state description in this theory consists of both ψ and $\mathbf{q}$. Bohmian mechanics does not involve a change in the mathematical form of ψ: just as in the quantum formalism, ψ is a vector in the Hilbert space associated with the system, and it evolves with time according to the Schrödinger equation:

$$i\hbar \frac{\partial \psi}{\partial t} = H\psi. \qquad (2.15)$$

The system configuration $\mathbf{q}$ is governed by the equation:

$$\frac{d\mathbf{q}}{dt} = (\hbar/m)\mathrm{Im}\left(\frac{\psi^* \nabla \psi}{\psi^* \psi}\right). \qquad (2.16)$$

In the case of a particle with spin, we make use of the spinor inner-product in this equation. For example, in the case of a spin $\frac{1}{2}$ particle whose wave function is $\psi = \chi_+(\mathbf{r})|\uparrow\rangle + \chi_-(\mathbf{r})|\downarrow\rangle$, the Eq. 2.16 assumes the form

---

[18] The term added to the particle's Hamiltonian to account for a magnetic field is $g\mathbf{s} \cdot \mathbf{B}$, where $\mathbf{s}$ is the spin, $\mathbf{B}$ is the magnetic field and g is the gyromagnetic ratio. To determine the form of this term in the case of a Stern–Gerlach apparatus, we require the configuration of the magnetic field. A Stern–Gerlach magnet apparatus has a "long axis" which for the example of Fig. 2.1 lies along the x-axis. Since the component of the magnetic field along this axis will *vanish* except within a small region before and after the apparatus, the effects of B_x may be neglected. Furthermore, B_y and B_z within the apparatus may be regarded as being independent of x. The magnetic field in the x, z plane between the magnets lies in the z-direction, i.e., $\mathbf{B}(x, 0, z) = B_z(z)\hat{k}$. Over the region of incidence of the particle, the field is such that $\frac{\partial B_z}{\partial z}$ is constant. See for example, Weidner and Sells [19] for a more detailed discussion of the Stern–Gerlach apparatus. The motion of the particle in the y-direction is of no importance to us, and so we do not consider the effects of any Hamiltonian terms involving only y dependence. The results we discuss in the present section are those which arise from taking account of the magnetic field by adding to the Hamiltonian term of the form $g\sigma_z B_z(z = 0) + g\sigma_z \left(\frac{\partial B_z}{\partial z}(z = 0)\right) z$.

$$\frac{d\mathbf{q}}{dt} = (\hbar/m)\text{Im}\left(\frac{\chi_+^*(\mathbf{r})\nabla\chi_+(\mathbf{r}) + \chi_-^*(\mathbf{r})\nabla\chi_-(\mathbf{r})}{\chi_+^*(\mathbf{r})\chi_+(\mathbf{r}) + \chi_-^*(\mathbf{r})\chi_-(\mathbf{r})}\right). \tag{2.17}$$

As we expect from the fact that this theory is in empirical agreement with quantum theory, Bohmian mechanics does *not* generally provide, given ψ and $\mathbf{q}$, a mapping from the observables to their values. In other words, it does not provide a non-contextual value map for each state. As we shall see, the choice of *experimental procedure* plays such a pronounced role in the Bohmian mechanics description of Albert's spin measurements, that one cannot possibly regard the spin operator as representative of an objective property of the particle.

We first discuss the Bohmian mechanics description of the Albert experiments. Since the wave function and its evolution are the same as in quantum mechanics, the particle's wave function ψ is taken to be exactly as described above. As far as the configuration $\mathbf{q}$ is concerned, there are two important features of the Bohmian evolution equations to be considered: the uniqueness of the trajectories and the *equivariance* of the time evolution. The first feature refers to the fact that each initial ψ and $\mathbf{q}$ leads to a *unique* trajectory. Since the particle being measured has a fixed initial wave function, its initial conditions are defined solely by its initial position. The equivariance of the system's time evolution is a more complex property. Suppose that at some time t, the probability that the system's configuration is within the region $d\mathbf{q}$ about $\mathbf{q}$ obeys the relationship:

$$P(\mathbf{q}' \in d\mathbf{q}) = |\psi(\mathbf{q})|^2 d\mathbf{q}. \tag{2.18}$$

According to equivariance, this relationship will continue to hold for all later times $t' : t' > t$. In considering the Bohmian mechanics description of a system, we assume that the particle initially obeys (2.18). By equivariance, we then have that for all later times the particle will be guided to "follow" the motion of the wave function. Thus, after it passes through the Stern–Gerlach apparatus, the particle will enter either the upward or downward moving packet. From consideration of the uniqueness of the trajectory and the equivariance of the time evolution, it follows that the question of *which branch* of the wave function the particle enters *depends solely on its initial position*.

If we consider the situation in a little more detail, we find a simple criterion on the initial position of the particle that determines which branch of the wave function it will enter. Recall that the initial (2.13) and final (2.14) wave function have no dependence on the y coordinate, and that they exhibit reflection symmetry through the $z = 0$ plane. From this symmetry together with the uniqueness of Bohmian trajectories, it follows that the particle cannot cross the $z = 0$ plane. In conjunction with equivariance, this result implies that if the particle's initial z coordinate is greater than zero, it must enter the upper branch, and if its initial z is less than zero the particle must enter the lower branch.

If we now consider the above described spin measurements, we find a somewhat curious situation. For any given initial configuration $\mathbf{q}$, the question of whether the measurement result is $\sigma_z = +\frac{1}{2}$ or $\sigma_z = -\frac{1}{2}$ depends on the configuration of the experimental apparatus. Suppose that the particle has an initial $z > 0$, so that according to the results just shown, it will enter the upper branch of the wave function. If the magnetic field inside the Stern–Gerlach apparatus is such that $\partial B_z/\partial z <$ 0, then the particle's final wave function is given by the first equation in (2.14), and its detection in the upper branch then implies that $\sigma_z = +\frac{1}{2}$. If, on the other hand, the magnetic field of the Stern–Gerlach magnet has the *opposite* orientation, i.e., $\partial B_z/\partial z > 0$, then the second equation in (2.14) obtains and the detection in the upper branch implies that $\sigma_z = -\frac{1}{2}$. Thus, we arrive at the conclusion that the "measurement of σ_z" gives a *different* result for two situations that differ only in the experimental configuration.

The quantum formalism's rules of measurement strongly suggest that the Hermitian operators represent objective properties of the system. Moreover, such a conception is a common element of the expositions given in quantum mechanics textbooks. On the other hand, the fact that the result of the "measurement" of σ_z can depend on properties of *both* system *and* apparatus contradicts this conception. In general, one must consider the results of the "measurement of an observable" to be a joint-product of system and measuring apparatus. Recall Niels Bohr's comment that [13] "a closer examination reveals that the procedure of measurement has an essential influence on the conditions on which the very definition of the physical quantities in question rests." For further discussion of the role of Hermitian operators in quantum theory, the reader is directed to Daumer, Dürr, Goldstein, and Zanghí in [17]. According to these authors: *"the basic problem with quantum theory . . . more fundamental than the measurement problem and all the rest, is a naive realism about operators . . . by (this) we refer to various, not entirely sharply defined, ways of taking too seriously the notion of operator-as-observable, and in particular to the all too casual talk about 'measuring operators' which tends to occur as soon as a physicist enters quantum mode."*

References

1. Gleason, A.M.: Measures on the closed subspaces of a Hilbert space. J. Math. Mech. **6**, 885–893 (1957)
2. Kochen, S., Specker, E.P.: The problem of hidden variables in quantum mechanics. J. Math. Mech. **17**, 59 (1967)
3. Mermin, N.D.: Simple unified form for the major no-hidden-variables theorems. Phys. Rev. Lett. **65**, 3373 (1990)
4. Pagels, H.R.: The Cosmic Code. Simon and Schuster, New York (1982)
5. Bell, J.S.: On the problem of hidden variables in quantum mechanics. Rev. Mod. Phys. **38**, 447–452 (1966). Reprinted in [21, p. 1] and [22, p. 397]
6. Bell, J.S.: On the impossible pilot wave. Found. Phys. **12**, 989–999 (1982). Reprinted in [1, p. 159]
7. Albert, D.: Quantum Mechanics and Experience. Harvard University Press, Cambridge (1992)
8. Belinfante, F.J.: A Survey of Hidden-Variables Theories. Pergamon Press, New York (1973)
9. Peres, A.: Incompatible results of quantum measurements. Phys. Lett. A **151**, 107 (1990)

10. Mermin, N.D.: Hidden variables and the two theorems of John Bell. Rev. Mod. Phys. **65**, 803 (1993)
11. Peres, A.: Quantum Theory: Concepts and Methods. Springer, Heidelberg (1995)
12. Bohm, D.: A suggested interpretation of the quantum theory in terms of "Hidden" variables. Phys. Rev. **85**, 166, 180 (1952)
13. Bohr, N.: Quantum mechanics and physical reality. Nature **136**, 65 (1935). Reprinted in [22, p. 144]
14. Schilpp, P.A.: Albert Einstein: Philosopher-Scientist. Harper and Row, New York (1949)
15. Kittel, C.: Introduction to Solid State Physics. 7th edn., Wiley, New York (1996)
16. Ghirardi, G.C.: Sneaking a Look at God's Cards. Princeton University Press, Princeton (2005)
17. Daumer, M., Düurr, D, Goldstein, S., Zanghí, N.: Naive realism about operators. In: Erkenntnis, 1996, Special issue in honor of Prof. Jeffrey, R., Costantini, D., Gallavotti, M.C. (eds.) Proceedings of the International Conference "Probability, Dynamics and Causality", Luino, Italy, 15–17 June 1995.
18. Bell, J.S.: Against 'Measurement'. Phys. World **3**, 33 (1990). This is reprinted in [23, p. 902]
19. Weidner, R.T., Sells, R.L.: Elementary Modern Physics, 2nd edn., Allyn and Bacon, Boston (1980)
20. Dürr, D., Goldstein, S., Zanghí, N.: Quantum equilibrium and the origin of absolute uncertainty. J. Statis. Phys. **67**, 843–907 (1992)
21. Bell, J.S.: Speakable and Unspeakable in Quantum Mechanics. Cambridge University Press, Cambridge (1987). (1987) Many of the works by Bell which are of concern to us may be found in this reference. See also [23] and [24]. The latter two are complete collections containing all of Bell's papers on quantum foundations
22. Wheeler, J.A., Zurek, W.H. (eds.): Quantum Theory and Measurement. Princeton University Press, Princeton (1983)
23. Bell, M., Gottfried, K., Veltman, M. (eds.): Bell on the Foundations of Quantum Mechanics, High Energy Physics and Accelerators: Selected Papers of John S. Bell (with commentary). World Scientific Publishing Company, Singapore (1995)
24. Bell, M., Gottfried, K., Veltman, M. (eds.): Bell on the Foundations of Quantum Mechanics. World Scientific Publishing Company, Singapore (2001)

Chapter 3
The Einstein–Podolsky–Rosen Paradox, Bell's Theorem and Nonlocality

3.1 Introductory Comments

In this chapter we turn to the paradox of Einstein, Podolsky and Rosen, [1] and Bell's Theorem[1] [11]. We wish to thoroughly review these matters, since the arguments of EPR and Bell run parallel to the reasoning of the Schrödinger paradox, which shall be the subject of the next chapter.

Bell's Theorem addresses issues not found in von Neumann's, Gleason's and Kochen and Specker's arguments. Bell's purpose in developing the theorem was not to refute hidden variables—for his part, Bell regarded the restoration of objectivity in quantum physics as a very worthwhile aim. Bell was quite impressed[2] with the success of David Bohm's objective quantum theory,[3] but was concerned about the presence of nonlocality. He wished to perform a test using a more abstract realistic formulation, hoping to verify whether nonlocality must be exhibited within every possible articulation of such a theory. To come to grips with this, Bell turned to the famous analysis known as the Einstein–Podolsky–Rosen paradox[4] and derived the theorem bearing his name. With this celebrated work, Bell had proven that nonlocality featured in Bohm's approach cannot be avoided in any objective theoretical formulation.

However, there is much more at play than just this. In addition, Bell and others have argued[5] that when the theorem is taken together with the Einstein–Podolsky–

---

[1] The account presented here follows similar lines as that of J.S. Bell and many other authors. See Bell in [2, 3]. Others authors who concur include Dürr, Goldstein, Zanghí (Sect. 8) in [4], Maudlin in [5, 6], Norsen [7–9] and Wiseman [10].

[2] See [12–16].

[3] See [17].

[4] To be precise, Bell worked directly with the David Bohm's spin-singlet version of Einstein–Podolsky–Rosen [18], pp. 611–623. A recent reprint appears within [19], pp. 356–368.

[5] See footnote 1 above. Not all authors agree with this position. Some reject quantum nonlocality. See Jarrett [20] and also Evans, Price and Wharton in [21]. Others argue that the theorem constitutes

D. L. Hemmick and A. M. Shakur, *Bell's Theorem and Quantum Realism*,
SpringerBriefs in Physics, DOI: 10.1007/978-3-642-23468-2_3,
© The Author(s) 2012

Rosen paradox, what follows is a stronger conclusion. Not only do we have a conflict of quantum mechanics with local realism, but Bell's Theorem implies the incompatibility of locality itself with quantum theory. I.e., one is left to conclude that quantum theory is irreducibly nonlocal.

In this chapter, we review and elaborate on these arguments. We attempt to present the issues in a very gradual manner, so that the essential nature of each step may be easily grasped. We present first the Einstein–Podolsky–Rosen paradox and Bell's Theorem individually, and then consider the implications of their conjunction. It is worth stressing here that the full ramifications of this chapter *follow from this combination* which we present in the final section of the chapter.[6]

3.2 Review of the Einstein–Podolsky–Rosen Paradox

3.2.1 Rotational Invariance of the Spin Singlet State and Perfect Correlations

The well-known work of Einstein, Podolsky, and Rosen, first published in 1935, was not designed to address the possibility of nonlocality, as such. The title of the paper was "Can Quantum Mechanical Description of Physical Reality be Considered Complete?", and the goal of these authors was essentially the opposite of authors such as von Neumann: Einstein, Podolsky, and Rosen wished to demonstrate that the addition of hidden variables to the description of state is *necessary* for a complete description of a quantum system. According to these authors, the quantum mechanical state description given by ψ is *incomplete*, i.e., it cannot account for all the objective properties of the system. This conclusion is stated in the paper's closing remark: "While we have thus shown that the wave function does not provide a complete description of the physical reality, we left open the question of whether such a description exists. We believe, however, that such a theory is possible."

Einstein, Podolsky, and Rosen arrived at this conclusion having shown that for the system they considered, each of the particles must have position and momentum as simultaneous "elements of reality". Regarding the completeness of a physical theory, the authors state: [1] (emphasis due to EPR) "Whatever the meaning assigned to the term complete, the following requirement for a complete theory seems to be a necessary one: every element of the physical reality must have a counterpart in

---

Footnote 5 (Continued)

a proof that realism is impossible in quantum physics. See Bethe [22], Gell-Mann [23], p. 172, and Wigner [19], p. 291.

[6] In particular, one cannot take the conclusion of the EPR paradox—the existence of noncontextual hidden variables—as received and final. This conclusion is based not only on quantum mechanical predictions, but also on the assumption of *locality*, which will ultimately be seen to fail. The status of EPR becomes clearer when one recognizes that the analysis is in fact equivalent to a theorem, as we demonstrate in Sect. 3.2.3.

the physical theory". This requirement leads them to conclude that the quantum theory is incomplete, since it does not account for the possibility of position and momentum as being simultaneous elements of reality. To develop this conclusion for the position and momentum of a particle, the authors make use of the following "sufficient condition" for a physical quantity to be considered as an element of reality: "If without in any way disturbing a system, we can predict with certainty (i.e. with probability equaling unity) the value of a physical quantity, then there exists an element of physical reality corresponding to this quantity".

What we shall present here[7] is a form of the EPR paradox which was developed in 1951, by David Bohm. Bohm's EPR analysis involves the properties of the *spin singlet state* of a pair of spin-$\frac{1}{2}$ particles. Within his argument, Bohm shows that various components of the spin of a pair of particles must be elements of reality in the same sense as the position and momentum were for Einstein, Podolsky, and Rosen. We begin with a discussion of the formal properties of the spin singlet state, and then proceed with our presentation of the EPR incompleteness argument.

Because the spin and its components all commute with the observables associated with the system's spatial properties, one may analyze a particle with spin by separately analyzing the spin observables and the spatial observables. The spin observables may be analyzed in terms of a Hilbert space $\mathcal{H}_s$ (which is a two-dimensional space in the case of a spin $\frac{1}{2}$ particle) and the spatial observables in terms of $L_2(\mathbb{R})$. The full Hilbert space of the system is then given by *tensor product* $\mathcal{H}_s \otimes L_2(\mathbb{R})$ of these spaces. Hence, we proceed to discuss the spin observables only, without explicit reference to the spatial observables of the system. We denote each direction in space by its θ and ϕ coordinates in spherical polar coordinates, and (since we consider spin $\frac{1}{2}$ particles) the symbol σ denotes the spin. To represent the eigenvectors of $\sigma_{\theta,\phi}$ corresponding to the eigenvalues $+\frac{1}{2}$ and $-\frac{1}{2}$, we write $|\uparrow \theta, \phi\rangle$, and $|\downarrow \theta, \phi\rangle$, respectively. For the eigenvectors of σ_z, we write simply $|\uparrow\rangle$ and $|\downarrow\rangle$. Often, vectors and observables are expressed in terms of the basis formed by the eigenvectors of σ_z. The vectors $|\uparrow \theta, \phi\rangle$, and $|\downarrow \theta, \phi\rangle$ when expressed in terms of these are

$$|\uparrow \theta, \phi\rangle = \cos(\theta/2)|\uparrow\rangle + \sin(\theta/2)e^{i\phi}|\downarrow\rangle$$
$$|\downarrow \theta, \phi\rangle = \sin(\theta/2)e^{-i\phi}|\uparrow\rangle - \cos(\theta/2)|\downarrow\rangle. \tag{3.1}$$

For a system consisting of *two* spin $\frac{1}{2}$ particles,[8] the states are often classified in terms of the total spin $S = \sigma^{(1)} + \sigma^{(2)}$ of the particles, where $\sigma^{(1)}$ is the spin of particle 1 and $\sigma^{(2)}$ is the spin of particle 2. The *spin singlet state*, in which we shall be interested, is characterized by $S = 0$. The name given to the state reflects that it contains just one eigenvector of the z-component S_z of the total spin: that corresponding to the eigenvalue 0. In fact, as we shall demonstrate, the spin singlet state is an eigenvector of *all* components of the total spin with an eigenvalue

---

[7] The Bohm spin singlet version and the original version of the EPR paradox differ essentially in the states and observables with which they are concerned. We shall consider the original EPR state more explicitly in Sect. 4.2.1.

[8] See for example, Messiah [24], and Shankar [25].

0. We express[9] the state in terms of the eigenvectors $|\uparrow\rangle^{(1)}, |\downarrow\rangle^{(1)}$ of $\sigma_z^{(1)}$ and $|\uparrow\rangle^{(2)}, |\downarrow\rangle^{(2)}$ of $\sigma_z^{(2)}$, as follows:

$$\psi_{ss} = |\uparrow\rangle^{(1)}|\downarrow\rangle^{(2)} - |\downarrow\rangle^{(1)}|\uparrow\rangle^{(2)}. \tag{3.2}$$

For simplicity, we have suppressed the normalization factor $\frac{1}{\sqrt{2}}$. Note that each of the two terms consists of a product of an eigenvector of $\sigma_z^{(1)}$ with an eigenvector of $\sigma_z^{(2)}$ such that the corresponding eigenvalues are the negatives of each other.

If we invert the relationships (3.1) we may then rewrite the spin singlet state (3.2) in terms of the eigenvectors $|\uparrow \theta, \phi\rangle$ and $|\downarrow \theta, \phi\rangle$ of the component of spin in the θ, ϕ direction. Inverting (3.1) gives:

$$|\uparrow\rangle = \cos(\theta/2)|\uparrow \theta, \phi\rangle + \sin(\theta/2)e^{i\phi}|\downarrow \theta, \phi\rangle$$
$$|\downarrow\rangle = \sin(\theta/2)e^{-i\phi}|\uparrow \theta, \phi\rangle - \cos(\theta/2)|\downarrow \theta, \phi\rangle. \tag{3.3}$$

Now let us consider rewriting the spin singlet state (3.2) in the following manner. We substitute for $|\uparrow\rangle^{(2)}$ and $|\downarrow\rangle^{(2)}$ expressions involving $|\uparrow \theta_2, \phi_2\rangle^{(2)}$ and $|\downarrow \theta_2, \phi_2\rangle^{(2)}$ by making use of (3.3). Doing so gives us a new expression for the spin singlet state:

$$\psi_{ss} = |\uparrow\rangle^{(1)}\left[\sin(\theta_2/2)e^{-i\phi_2}|\uparrow \theta_2, \phi_2\rangle^{(2)} - \cos(\theta_2/2)|\downarrow \theta_2, \phi_2\rangle^{(2)}\right]$$
$$- |\downarrow\rangle^{(1)}\left[\cos(\theta_2/2)|\uparrow \theta_2, \phi_2\rangle^{(2)} + \sin(\theta_2/2)e^{i\phi_2}|\downarrow \theta_2, \phi_2\rangle^{(2)}\right]$$
$$= -\left[\cos(\theta_2/2)|\uparrow\rangle^{(1)} + \sin(\theta_2/2)e^{i\phi_2}|\downarrow\rangle^{(1)}\right]|\downarrow \theta_2, \phi_2\rangle^{(2)}$$
$$+ \left[\sin(\theta_2/2)e^{-i\phi_2}|\uparrow\rangle^{(1)} - \cos(\theta_2/2)|\downarrow\rangle^{(1)}\right]|\uparrow \theta_2, \phi_2\rangle^{(2)}. \tag{3.4}$$

Examining the second of these relationships, we see from (3.1) that ψ_{ss} reduces[10]:

$$\psi_{ss} = |\uparrow \theta_2, \phi_2\rangle^{(1)}|\downarrow \theta_2, \phi_2\rangle^{(2)} - |\downarrow \theta_2, \phi_2\rangle^{(1)}|\uparrow \theta_2, \phi_2\rangle^{(2)}. \tag{3.5}$$

Note that this form is similar to that given in (3.2). Each of its terms is a product of an eigenvector of $\sigma_{\theta_2,\phi_2}^{(1)}$ and an eigenvector of $\sigma_{\theta_2,\phi_2}^{(2)}$ such that the factors making up the product correspond to eigenvalues that are just the opposites of each other. We drop the '2' from θ and ϕ giving

$$\psi_{ss} = |\uparrow \theta, \phi\rangle^{(1)}|\downarrow \theta, \phi\rangle^{(2)} - |\downarrow \theta, \phi\rangle^{(1)}|\uparrow \theta, \phi\rangle^{(2)}. \tag{3.6}$$

---

[9] Note that a term such as $|a\rangle^{(1)}|b\rangle^{(2)}$ represents a *tensor product* of the vector $|a\rangle^{(1)}$ of the Hilbert space associated with the first particle with the vector $|b\rangle^{(2)}$ of the Hilbert space associated with the second. The formal way of writing such a quantity is as: $|a\rangle^{(1)} \otimes |b\rangle^{(2)}$. For simplicity of expression, we shall omit the symbol '$\otimes$' here.

[10] If we multiply a wave function by any constant factor c, where $c \neq 0$ the resulting wave function represents the same physical state. We multiply ψ_{ss} by -1 to facilitate comparison with (3.2).

Suppose that we consider the implications of (3.6) for individual measurements of $\sigma_{\theta,\phi}^{(1)}$ of particle 1 and the same spin component of particle 2. In each term of this spin singlet form, we have a product of eigenvectors of the two spin components such that the eigenvalues are simply the negatives of one another. Therefore, it follows that *measurements of $\sigma_{\theta,\phi}^{(1)}$ and $\sigma_{\theta,\phi}^{(2)}$ always give results that sum to zero*, i.e. if measurement of $\sigma_{\theta,\phi}^{(1)}$ gives the result $\pm\frac{1}{2}$, then the measurement of $\sigma_{\theta,\phi}^{(2)}$ always gives $\mp\frac{1}{2}$. We say that these observables exhibit *perfect correlation*. Note that this holds true for pairs of spin observables $\sigma_{\theta,\phi}^{(1)}$ and $\sigma_{\theta,\phi}^{(2)}$ where θ, ϕ refers to a direction in space, as specified in spherical polar coordinates.

3.2.2 The EPR Incompleteness Argument and Objective Realism

Consider a situation in which the two particles described by the spin singlet state are spatially separated from one another, and spin-component measurements are to be carried out on each. Since there exist perfect correlations, it is possible to predict with certainty the result of a measurement of $\sigma_x^{(1)}$ from the result of a previous measurement of $\sigma_x^{(2)}$. Suppose for example, we measure $\sigma_x^{(2)}$ and find the result $\sigma_x^{(2)} = \frac{1}{2}$. The subsequent measurement of $\sigma_x^{(1)}$ must give $\sigma_x^{(1)} = -\frac{1}{2}$. Of course, it is also the case that the measurement result $\sigma_x^{(2)} = -\frac{1}{2}$ allows prediction that an appropriate ensuing measurement on particle 1 must give $\sigma_x^{(1)} = \frac{1}{2}$. If we assume *locality* then the measurement of $\sigma_x^{(2)}$ cannot in any way disturb particle 1, which is spatially separated from particle 2. Using the Einstein–Podolsky–Rosen criterion, it follows that $\sigma_x^{(1)}$ is an element of reality.

3.2.2.1 Physical Implications of EPR and Non-contextuality

The above argument is brief and concise, employing straightforward concepts. However, we cannot be satisfied with just this. The physical state of affairs of the described scenario is important both in itself and as it relates to Bell's Theorem.

As we have seen, EPR employed the perfect predictability of the spin-component measurement to develop somewhat dramatic consequences. Evidently the argument suggests that for an ensemble of such systems, one may perform arbitrarily many trials and the above mentioned predictability will never fail. Such perfect prediction implies that some matter of fact must obtain for the value of the observable $\sigma_x^{(1)}$.[11]

---

[11] Already we see a contrast with the point of view of quantum theory, which asserts that no physical property has meaning apart from a measurement procedure. One could at this point assert that the incompleteness of quantum mechanics has been established. It is interesting to note

Thus, the EPR argument has established that (assuming locality) there must exist a distinct value for the observable. However, the formal rules of quantum mechanics and the state representation given by ψ make no allowance for the existence of the observable $\sigma_x^{(1)}$ and its value, save for those particular times after the measurement process has been performed. Mathematical representation of the value must be in terms of some new parameter. We shall refer to the new parameter as λ, for consistency with Sect. 1.4.1. Using the notation of that section, the Einstein–Podolsky–Rosen paradox implies that there must exist a value-map function $V_\lambda(\sigma_x^{(1)})$.[12]

Note well that the value map $V_\lambda(\sigma_x^{(1)})$ makes no allowance for the choice of measurement procedure which might be employed in confirming the value. The reason is that within the EPR paradox, the perfect correlations—on which the element of reality condition depends—are themselves independent of which measurement options are selected.[13]

3.2.2.2 EPR Conclusion and Hidden Variables

Now let us return to the EPR argument itself, and examine the case for some other component of particle 1's spin, $\sigma_{\theta,\phi}^{(1)}$. In fact, just as was the case for $\sigma_x^{(1)}$, one can predict with certainty the result of a measurement of $\sigma_{\theta,\phi}^{(1)}$ from a previous measurement of $\sigma_{\theta,\phi}^{(2)}$. Again, locality implies that measurement of $\sigma_{\theta,\phi}^{(2)}$ does not disturb particle 1, and by the EPR criterion, $\sigma_{\theta,\phi}^{(1)}$ must be an element of reality. Hence, particle 1's spin component in an arbitrary direction θ,ϕ must be an element of reality.

As in the case of $\sigma_x^{(1)}$ discussed above, the formalism of quantum mechanics cannot account for the existence of predetermined values on the spin components $\sigma_{\theta,\phi}^{(1)}$. As we extend the quantum mechanical state description and supplement ψ with the parameter λ, we can now mathematically represent the conclusion of the EPR paradox thus far by a value-map function $V_\lambda(\sigma_{\theta,\phi}^{(1)})$. Two remarks are in order regarding this form. As in the above case for $\sigma_x^{(1)}$, there is no dependence upon the

---

Footnote 11 (Continued)
in this connection that Einstein [26, pp. 167–168] disliked that Podolsky and Rosen had formulated incompleteness by reference to both x and p when just one quantity would suffice.

[12]   A few things may need to be clarified. First, no dependence on ψ need be included in this formulation since we are considering a fixed wave-function, namely that of the spin-singlet state. Second, writing such a mathematical function does not in any sense constitute an additional assumption. The existence of particular values which match measurement results perfectly assures that a value map of just this form must exist.

[13]   Readers may object that on the relevant spin space each component of the spin-$\frac{1}{2}$ observable is non-degenerate, so that the considerations of measurement procedure play very little role. However, an observable need not be a member of a commuting family in order for a choice of measurement procedures to come into play. Even a spin-$\frac{1}{2}$ particle exhibits contextuality and dependence of measurement result upon the choice of experimental procedure. See Sect. 2.5.

decision of measurement procedure which might subsequently be taken to test the value map prediction. The perfect correlations are simply not sensitive to this aspect of the operations. Secondly, if we consider a single member of an ensemble of spin singlet states, i.e., a fixed value of λ, then this value map implies a mapping of the form $E(\sigma^{(1)}_{\theta,\phi})$ which takes the spin components to their values. Furthermore, the independence of measurement procedure means that E is a non-contextual mapping of just the sort considered by Gleason and also by Kochen and Specker.

Note that it is not difficult at this point to establish the incompleteness of quantum theory, as EPR famously contended. Quantum theory does not permit the assignment of simultaneous values to incompatible physical quantities. Therefore, the argument's conclusion implies that this theory is incomplete, since such values have been derived.[14]

As a final step, note that we can interchange the roles of particles 1 and 2 in this argument to show the same conclusion for all components of particle 2's spin, as well.

Every component of spin of both particles is therefore an element of reality. If there is a particular matter of fact about these spin components, then we arrive at the same conclusions just given for components not only of particle one, but for both particles. The conclusion of the Einstein–Podolsky–Rosen paradox[15] is the existence of non–contextual hidden variables on the spin components of the two particles.

3.2.3 The Einstein–Podolsky–Rosen Theorem

While the argument of the Einstein–Podolsky–Rosen paradox is not an overly complicated one, it is important to be certain each element is clearly enumerated and understood. Doing so leads us to an important result, as we will find that EPR's analysis is equivalent to a theorem.

The concept of an element of reality evidently plays a central role. As discussed above Einstein–Podolsky–Rosen regard a sufficient condition for an element of reality to be the following: [1] "If without in any way disturbing a system, we can predict with certainty (i.e., with probability equal to unity) the value of a physical quantity, then there exists an element of physical reality corresponding to this quantity." This condition turns out to be precisely what is needed in the argument of EPR.

The sufficient condition is thus based upon two criteria. The first arises from the phrase "without in any way disturbing a system." Going back to the argument, it is clear that the spatial separation of the two particles is what allows this prerequisite

---

[14] Among the set $\{\sigma^{(1)}_{\theta,\phi}\}$ it is of course, trivial to locate a pair of incompatible observables, e.g., $\sigma^{(1)}_x$ and $\sigma^{(1)}_y$.

[15] Once again, we emphasize that this conclusion ought not be taken as the final word on the spin singlet state. It follows *only* if one admits the assumption of locality, which axiom is to meet its doom, once we are aware of Bell's Theorem and its implications. The death of locality is to to discussed in the final section of the chapter.

to be fulfilled. It is presumed that because the particles are separated, there can be no immediate influence of one upon the other.

Upon reflection it becomes clear that the assertion just made, that spatial separation implies freedom from influence, rests squarely on the assumption of locality. To drop the notion of locality would be to allow mutual instantaneous influences between the particles, thereby destroying their independence. If we give up locality, we can no longer assert that the procedures are taking place "without in any way disturbing a system".

The second criterion for an element of reality is the "prediction with certainty." This type of prediction is just what we gain from the perfect correlations, as we have seen.

Because both criteria have been met, it follows from the above sufficient condition that each and every quantity which exhibits perfect correlations must be an element of reality. In particular, for each spin-component, there must exist distinct and unambiguous values which predetermine the measurement results.

The content of the Einstein–Podolsky–Rosen paradox is equivalent to the following logical structure. Given locality together with perfect correlations, we have non-contextual hidden variables on all spin components of the two particles represented by the spin-singlet state. Although tradition calls for use of the name "paradox,"[16] there is no reason not to refer to it as the "EPR theorem."

Having seen all this, it becomes obvious that the implications of EPR are broader than they might appear at first glance. Certainly one could conceive of any number of hypothetical theories which would at once exhibit such perfect correlations as well as being of a local character. Because the EPR theorem demands only these features, a variety of theories fall under its purview.

Such broad applicability of the EPR analysis will prove itself crucial when we come to consider the Bell's theorem and its ramifications. We now turn to this topic.

3.3 Bell's Theorem

To begin the discussion, we first modify the notation somewhat from that utilized above. To denote directions in space, we write unit vectors such as $\hat{a}$, $\hat{b}$, $\hat{c}$ instead of θ and ϕ. Rather than using the form $V_\lambda(O)$, we shall write $A(\lambda, \hat{a})$ and $B(\lambda, \hat{b})$ to represent functions on the spin components of particles 1 and 2, respectively. Since the two particles are each of spin $\frac{1}{2}$, we should have $A = \pm\frac{1}{2}$ and $B = \pm\frac{1}{2}$, however for simplicity we rescale these to $A = \pm1$ and $B = \pm1$.

---

[16] As everyone knows, a paradox is a self-contradiction. However, this name is not descriptive of what the analysis offers. In the situation analyzed by EPR, it happens that the conclusion runs counter to quantum mechanics. Here, quantum mechanics is essentially being used to point the road to its own limitations.

3.3.1 Proof of Bell's Theorem

The key feature of the spin singlet version of the EPR paradox was its analysis of the perfect correlations arising when the two particles of a spin singlet pair are subject to measurements of the same spin component. Thus, it may not be surprising that Bell's Theorem is concerned with a *correlation function*, which is essentially a measure of the statistical correlation between the results of spin component measurements of the two particles. The correlation function is to be determined as follows: we set the apparatus measuring particle 1 to probe the component in the $\hat{a}$ direction, and the apparatus measuring 2 is set for the $\hat{b}$ direction. We make a series of measurements of spin singlet pairs using this configuration, recording the *product* $\sigma_{\hat{a}}^{(1)}\sigma_{\hat{b}}^{(2)}$ of the results on each trial. The average of these products over the series of measurements is the value of the correlation function.

In general, we expect the value of the average determined in this way to depend on the directions $\hat{a}$, $\hat{b}$ with respect to which the spin components are measured. According to the quantum formalism, we may predict the average, or *expectation value* of any observable using the formula $E(O) = \langle \psi | O \psi \rangle$. For the series of experiments just described, we take the expectation value of product of the appropriate spin component observables, giving:

$$P_{QM}(\hat{a}, \hat{b}) = \langle \sigma_a^{(1)}\sigma_b^{(2)} \rangle = -\hat{a} \cdot \hat{b}. \tag{3.7}$$

In the case of the predetermined values, the average of the product of the two spin components $\sigma_{\hat{a}}^{(1)}\sigma_{\hat{b}}^{(2)}$ is obtained by taking an average over λ:

$$P(\hat{a}, \hat{b}) = \int d\lambda \rho(\lambda) A(\lambda, \hat{a}) B(\lambda, \hat{b}), \tag{3.8}$$

where $\rho(\lambda)$ is the probability distribution over λ. $\rho(\lambda)$ is normalized by:

$$\int d\lambda \rho(\lambda) = 1. \tag{3.9}$$

We will now examine the question of whether the correlation function given by (3.8) is compatible with the quantum mechanical prediction (3.7) for this function.

Crucial to the EPR analysis is the fact that there is a perfect correlation between the results of the measurement of any component of particle 1's spin in a given direction with the measurement of the same component of particle 2's spin, such that the results are of opposite sign. To account for this, the correlation function must give

$$P(\hat{a}, \hat{a}) = -1 \ \forall \hat{a}. \tag{3.10}$$

It is easy to see that the quantum correlation function satisfies this condition. If the prediction derivable using the predetermined values is to reflect this, we must have

$$A(\lambda, \hat{a}) = -B(\lambda, \hat{a}) \; \forall \hat{a}, \lambda. \tag{3.11}$$

At this point, we have enough information to derive the conclusion of the theorem. Using (3.8) together with (3.11) and the fact that $[A(\lambda, \hat{a})]^2 = 1$, we write

$$P(\hat{a}, \hat{b}) - P(\hat{a}, \hat{c}) = -\int d\lambda \rho(\lambda)[A(\lambda, \hat{a})A(\lambda, \hat{b}) - A(\lambda, \hat{a})A(\lambda, \hat{c})]$$

$$= -\int d\lambda \rho(\lambda)A(\lambda, \hat{a})A(\lambda, \hat{b})[1 - A(\lambda, \hat{b})A(\lambda, \hat{c})] \tag{3.12}$$

Using $A, B = \pm 1$, we have that

$$|P(\hat{a}, \hat{b}) - P(\hat{a}, \hat{c})| \le \int d\lambda \rho(\lambda)[1 - A(\lambda, \hat{b})A(\lambda, \hat{c})]; \tag{3.13}$$

then using the normalization (3.9), and (3.11) we have

$$|P(\hat{a}, \hat{b}) - P(\hat{a}, \hat{c})| \le 1 + P(\hat{b}, \hat{c}), \tag{3.14}$$

and this relation, which is commonly referred to as "Bell's inequality", is the theorem's conclusion.

Thus, the general framework of Bell's Theorem is as follows. The definite values of the various components of the two particles' spins are represented by the mathematical functions $A(\lambda, \hat{a})$, and $B(\lambda, \hat{b})$. The condition

$$A(\lambda, \hat{a}) = -B(\lambda, \hat{a}) \quad \forall \hat{a}, \lambda \tag{3.15}$$

(Eq. 3.11), placed on the functions $A(\lambda, \hat{a})$, $B(\lambda, \hat{b})$ ensures the agreement of these functions with the perfect correlations. Bell's Theorem tells us that from these conditions it follows that the theoretical prediction for the correlation function, $P(\hat{a}, \hat{b})$, must satisfy the Bell inequality, (3.14).

Consider now the quantum mechanical prediction for the correlation function (3.7). If we examine this function, we find that it does *not* in general satisfy Bell's inequality. Suppose, for example, that we have defined some angular orientation such that $\hat{a}, \hat{b}, \hat{c}$ all lie in the x, y plane (so that $\theta = 90°$), with $\hat{a}$ along $\phi = 60°$, $\hat{b}$ along $\phi = 0°$, and $\hat{c}$ along $\phi = 120°$. With this, we have $P_{QM}(\hat{a}, \hat{b}) = \frac{1}{2}$, $P_{QM}(\hat{a}, \hat{c}) = \frac{1}{2}$ and $P_{QM}(\hat{b}, \hat{c}) = -\frac{1}{2}$, so that $|P_{QM}(\hat{a}, \hat{b}) - P_{QM}(\hat{a}, \hat{c})| = 1$ and $1 + P_{QM}(\hat{b}, \hat{c}) = \frac{1}{2}$, which is in violation of (3.14).

Therefore, according to Bell's Theorem, the existence of non-contextual hidden variables on the spin components leads to a contradiction with the quantum mechanical predictions.

3.4 Einstein–Podolsky–Rosen, Bell's Theorem, and Nonlocality

3.4.1 Complex Arguments

An argument is a series of statements including one or more premises together with a conclusion. If the argument is valid, then the conclusion follows necessarily from the premises. When one provides a demonstration of an argument's validity, then we say that we have proven a theorem.

The notion of a complex argument may be a more infrequent concept. Suppose we have two simple arguments. Suppose further that the conclusion of the first argument is identical to one of the second argument's premises. We might then imagine that the conclusion of the second argument is in turn the premise of a third argument, and so on. When one combines arguments like these and creates a cascading structure, the result is what is known as a complex argument.

Thus within a complex argument, there are statements which serve as "stepping stones"; i.e., those which are common to two of the component arguments. When one refers to a statement which acts as both a conclusion and as a premise, one uses either the term 'intermediate conclusion,' or 'nonbasic premise'. A nonbasic premise is in contrast to a 'basic premise', which in fact is *not* the conclusion of another argument.

If every stage of a complex argument is valid, then it follows that the argument as a whole is also valid. This is an intuitive result which is not difficult to see. First, the truth of the basic premises assures the truth of the intermediate conclusion which follows in the argument's first stage. Using the intermediate conclusion with the basic premises of the second stage then assures the truth of that argument's conclusion. The process may be continued step by step all the way to the point at which the truth of the final argument's conclusion is established.

3.4.2 Conjunction of EPR Analysis with Bell's Theorem

If we now consider[17] the Einstein–Podolsky–Rosen Theorem with Bell's Theorem it is not difficult to see that we have the right scenario to construct a complex argument. First recall the logic of the Einstein–Podolsky–Rosen Theorem. From the premises of perfect correlations and locality, it follows that the spin components are objective physical properties represented by non-contextual hidden variables. As we have remarked, the generality of these premises implies that the theorem is applicable to a broad class of theories. Bell's Theorem, as we saw, tells us that perfect

---

[17] We will use the term "the Einstein–Podolsky–Rosen Theorem" in this section, in spite of the fact that it is not a term which appears in the literature. We do so for ease of discussion, to avoid using cumbersome phrases such as "the theorem implied by the Einstein–Podolsky–Rosen paradox." The proof that the Einstein–Podolsky–Rosen paradox is equivalent to a theorem is presented in Sect. 3.2.3.

correlations together with non-contextual hidden variables on the spin components leads inexorably to the Bell inequality.

Since the conclusion of the Einstein–Podolsky–Rosen Theorem is also one of the premises of Bell's Theorem one may combine the two to form a complex argument, as discussed above. Within this complex argument we find that *the existence of hidden variables on the spins is a nonbasic premise of the Bell's Theorem component.*

The validity of this complex argument follows since its components are theorems. When we examine the theorems to collect the basic premises, we find that those of the Einstein–Podolsky–Rosen Theorem are locality and perfect correlations, while Bell's Theorem is based on perfect correlations.[18] The conclusion of the overall argument is just that of Bell's Theorem, namely the Bell inequality. In total, we have this structure: given locality together with perfect correlations, it follows that the Bell inequality must hold true.

Note carefully what is being said here. The argument assumes only locality and perfect correlations. Since these assumptions leave room for a spectrum of hypothetical theories, we have come to a powerful conclusion. When we consider the empirical predictions of any theoretical structure which meets these premises, such predictions must obey the Bell inequality. However, we saw above that the quantum predictions for the spin singlet state necessarily conflict with the Bell inequality.

Let us try to bring all this together, as we summarize the situation. For the sake of argument, let us suppose the quantum prediction of perfect correlations had proven themselves false in the laboratory. In that case, pursuing the Einstein–Podolsky–Rosen paradox and Bell's Theorem would be following a "false trail," and we could say that in its prediction of perfect correlations, quantum theory has broken down. If, on the other hand, the perfect correlations are true and accurate representations of laboratory behavior, an entirely different scenario results. In this case, from the results just obtained there are two distinct possibilities regarding the theoretical description. On the one hand, we have the quantum theory. On other other, we have the family of local physical theories, all of which obey Bell's inequality. *Because quantum mechanics violates Bell's inequality, it is in empirical disagreement with the family of local physical theories.*

Thus, we have that if the quantum predictions are correct, then there is no way to explain the experimental results using any local theory. And indeed, experiments have supported quantum mechanics in this regard,[19] so that we may conclude also that nature is nonlocal.

---

[18] The latter is due to the fact that what previously served as a premise of Bell—the existence of non-contextual hidden variables—turns out to be a *nonbasic* premise within the complex argument.

[19] The large majority of experiments have dealt with a two-photon system, rather than two spins. See for example, Freedman and Clauser [27], Fry and Thompson [28], and Aspect et al. [29–31]. For such quantum systems, the most definitive theoretical analysis was given by Clauser, Horne, Holt and Shimony [32]. Here the photon polarizations play the role of the spin components. The experiment of Lamehi-Rachti, and Mittig [33] involves a pair of protons described by the spin singlet state. For a discussion of quantum nonlocality experiments, see Bell in [2, 34] and Herbert in [35].

3.4.3 The 'No-Superdeterminism' Assumption

Bell's theorem strongly suggests the performance of an experimental program to decide between local theories and the quantum theory. As noted above, such experiments have been performed, with the results favoring the latter. However, in order for our discussion to be complete, it is necessary to mention an argument[20] some have applied for the purpose of avoiding the nonlocality conclusion. However, it is not difficult to see that the proposed idea—brought in to weaken or refute nonlocality—is not only artificial but quite radical and unwarranted, as we shall see.

To begin, we first consider the sort of correlation experiment which Bell's theorem calls for. One would like to perform a series of spin measurements on the two particles to resolve whether the results satisfy the quantum predicions or Bell's inequality. In this regard, one of the simplest considerations is the *timing* with which the two diligent scientists select the settings of their respective experiments.[21] Because the analysis of concern depends so crucially upon the notion of locality, one must be certain the measurements can truly be regarded as spacelike separated. Therefore, the choice of which spin component is tested by say, experimenter A must be made 'at the last minute' i.e., such that even a signal travelling at the speed of light could not inform the other particle of this choice. If the experimenters do not meet such a 'last minute choice' criterion, then one could argue that the measurement of particle-1 *does* disturb particle-2 so that the EPR element of reality criterion cannot be met.[22]

Suppose we consider a series of experiments which meets this sort of requirement and thereby eliminates the possibility of all subluminal signals which might "notify the right hand of what the left is doing." Is that the only requirement for the logic to carry through? In fact, there is another logical possibility which could account for the perfect correlations *without* forcing a choice between local theories and quantum mechanics, and this is the main point of the section.

In brief, one could call the hypothesis a kind of "superdeterminism," according to which each experimenter's selections for measurement are actually rigorously fixed in advance for every run of the series. In other words, while each experimenter certainly considers themselves free to measure whatever spin axis they wish, we are supposing that *no* such freedom actually exists, but that their belief is a kind of illusion. In addition, we might imagine that the parameter λ which characterizes the state is *also* so predetermined, and that this also holds true for each experimental run. Finally, let the time-progression of the scientists and quantum system be such

---

[20] Bell briefly confronts the concept in [36], p. 47. He discusses the interpretation that "the world is superdeterministic." In another work, [37, 38] Bell treats the same concept in a somewhat more mathematical and detailed manner. See also Goldstein, Norsen, Tausk, Zanghi in [39], specifically their reference to the 'no-conspiracies' assumption. We find the Bell term 'superdeterminism' more conducive to the expostulation of the principle involved.

[21] See for example, G.C. Ghirardi [40], pp. 243–246.

[22] Note that the *means* of such disturbance requires an inventive imagination. It seems rather odd to suppose that a signal might pass from the spin-measuring apparatus on one side of the lab to the particle on the other. However, since the question at issue is one that "shakes the foundations" of physical science, it might be reasonable to account for even such unknown and unusual hypotheses.

that at every trial of the experiment, the quantum system is in perfect lockstep with the experimenters' "selections" of measurement axes to analyze.[23]

For what its worth, appeal to such a consideration allows one to avoid altogether the issue of quantum nonlocality. If we permit superdeterminism of this kind, involving the rigorous establishment of both human experimenters and of the quantum system, then the EPR/Bell argument is essentially halted from the beginning. The freedom of each of these parameters is a tacit assumption of the thinking at work in the Einstein–Podolsky–Rosen paradox.

Two questions come to the fore at this point. First, the issue of superdeterminism appears quite closely related to human free will. How closely are the two linked? If in fact, we *deny* the absolute regulation of human and quantum behavior, is this tantamount to the assumption of human free will? The second question concerns the status of the Einstein–Podolsky–Rosen paradox and Bell's Theorem. Should we take the possibility of superdeterminism as a blow to quantum nonlocality?

Consider the first question. Upon reflection, it is clear that denial of superdeterminism is not identical to the assumption of free will. It is not difficult to see that such superdeterminism on the one hand, and the supposition of human free will on the other are opposite extremes along a continuum. There is a great amount of "daylight" between the two positions. Consider, for example, an experimenter faced with the choice[24] of two positions, say 45 degrees and 135 degrees for a classic Stern–Gerlach magnet. In the usual picture, it is natural to imagine no such restriction, but that one may choose any arbitrary angle. But let us suppose for the sake of argument that there are just two selections available. Then, on the superdeterminism picture, one would have to regard one position, say 135 degrees, as inescapable, and the other as impossible. In other words, in this picture, there is a 100 hundred per cent chance the experimenter will choose 135 and no chance of her selecting the other. But the free will assumption would imagine that nothing whatever governs the choice, so that we have a "fifty–fifty" distribution of odds over the two directions. Clearly, there is a large range between the two positions. One might imagine any number of ways that the experimenter is drawn away from perfect free will. The distraction of a person entering the room, the shaking of a hand reaching for the dial,[25] and so on. The fact that one cannot easily quantify such a thing seems clear, but it does *not* follow that such effects do not exist. It is also evident that these factors must detract from the ideal of perfect free will.

Thus, if we make the assumption that there is no superdeterminism, this is not comparable to a belief in human free will. Rather it simply constitutes a movement

---

[23] In a hidden variables context, this would mean that the quantum system has "preprogrammed answers" for the measurements which will be made by the two scientists. But Bell's theorem is derived based upon a continuously infinite set of hidden variables, and it is not clear whether or how Bell's conclusion—empirical disagreement between the different styles of theories—can be developed from a small and finite set.

[24] Or perhaps we could call it "pseudo-choice" if we take seriously the position of superdeterminism.

[25] To follow the example offered by Bell [33, 34].

towards the "middle ground" between the two extremes. It is clear that denying superdeterminism is a weaker assumption than that of perfect free will.

Having seen that the EPR/Bell argument rests also upon the no superdeterminism assumption as we have defined it, should we then consider the conclusion—quantum nonlocality—as somehow blocked or refuted? Evidently not, when one examines the reasonability of superdeterminism. On its surface, superdeterminism violates our general overall sense of things: we do not regard ourselves as forced to adhere to one and only one selection in every particular choice we might face.[26] But beyond this, the idea is radical enough to make meaningless the essential foundations of scientific activity.[27] A scientific investigator must have some amount of freedom to design and execute experiments which are free from bias. In short to assume that such superdeterminism does not hold true is not a strong assumption at all, but constitutes one of the most basic premises underlying all science.

Therefore, the assumption that there is no superdeterminism does essentially nothing to detract from the quantum nonlocality conclusion, but simply reaffirms one of the basic assumptions of science.

3.5 Summary

Bell's Theorem was developed as an effort to investigate whether nonlocality must be a necessary feature of realistic formulations of quantum theory. The result was of course, that all theories featuring local hidden variables must conflict with the quantum mechanical predictions. As compelling as this may be in itself, this is not yet the full story, as Bell showed in further papers.[28] Bell directed attention to the Einstein–Podolsky–Rosen paradox and its implications, and showed that by considering this argument in conjunction with his own theorem, a different result follows. What is implied by such a combination is that the empirical predictions of quantum mechanics are in conflict with any local theoretical explanation one might consider.[29] The essential point involved in the derivation of this result is that hidden variables constitute a *non-basic premise* when one views the conjunction of Einstein–Podolsky–Rosen Theorem and Bell's Theorem as a complex argument.

What can be said regarding the possibility of objective interpretations of quantum phenomena? While there is no contradicting the notion that local realistic theories

---

[26] Although there are some obvious restrictions set down by social norms and by biological limitations.

[27] Indeed, such a "martial law" precisely fixing human actions in lockstep with quantum states invokes a scenario as bizarre as any science fiction. Does this not remove the meaning most people ascribe to *all* human activities not just scientific, but intellectual, cultural and even social?

[28] See [2, 3] Also see Dürr, Goldstein, Zanghí (Sect. 8) in [4], Maudlin in [5, 6], Norsen [7] and Wiseman [10].

[29] As noted above, there is another assumption behind the logic here, namely what we called the 'no-superdeterminism'. However, this does not constitute a refutation of the argument, but only an explicit recognition of the same assumption that underlies all scientific research.

must be ruled out, we see also that the quantum theory itself conflicts with all local theoretical explanations. Quantum theory may be called *irreducibly* nonlocal. Therefore, the EPR/Bell analysis does not serve to distinguish quantum mechanics from a candidate hidden variables theory, but simply reveals that every viable formulation of quantum physics must exhibit such nonlocality.

References

1. Einstein, A., Podolsky, B., Rosen, N.: Can quantum-Mechanical description of physical reality be considered complete? Phys. Rev. **47**, 777 (1935). Reprinted in [19, p. 138]
2. Bell, J.S.: Atomic cascade photons and quantum mechanical non-locality. Comments At. Mol. Phys. **9**, 121–126 (1980). Reprinted in [41, p. 782]
3. Bell, J.S.: Bertlemann's socks and the nature of reality. Journal de Physique Colloque C2, suppl. au numero 3, Tome 42 1981, pp. C2 41–61 (1981). Reprinted in [42, p. 139]
4. Dürr, D., Goldstein, S., Zanghí, N.: Quantum equilibrium and the role of operators as observables in quantum theory. J. Stat. Phys. **116**, 959–1055 (2004)
5. Maudlin, T.: Quantum Non-Locality and Relativity: Metaphysical Intimations of Modern Physics, 3rd edn. Wiley-Blackwell, Oxford (2011)
6. Maudlin, T.: Space-time in the quantum world. In: Cushing, J., Fine, A., Goldstein, S., (eds.) Bohmian Mechanics and Quantum Theory: an Appraisal, pp. 285–307. Kluwer, Dordrecht (1996)
7. Norsen, T.: Against 'realism'. Found. Phys. **37**, 311–340 (2007)
8. Norsen, T.: Bell locality and the nonlocal character of nature. Found. Phys. Lett. **19**(7), 633–655 (2006)
9. Norsen, T.: Local causality and completeness: Bell vs. Jarrett. Found. Phys. **39**(3), 237 (2009)
10. Wiseman, H.M.: From Einstein's theorem to Bell's theorem: A history of quantum nonlocality. Contemp. Phys. **47**, 79–88 (2006)
11. Bell, J.S.: On the Einstein Podolsky Rosen paradox. Physics **1**, 195–200 (1964). Reprinted in [42, p. 14] and in [19, p. 403]
12. Bell, J.S.: Six possible worlds of quantum mechanics. In: Proceedings of the Nobel Symposium 65: Possible Worlds in Arts and Sciences. Stockholm, August 11–15 (1986). Reprinted in [42, p. 181]
13. Bell, J.S.: On the impossible pilot wave. Found. Phys. **12**, 989–999 (1982). Reprinted in [42, p. 159]
14. Bell, J.S.: Quantum mechanics for cosmologists. In: Isham, C., Penrose, R., Sciama, D. (eds.) Quantum Gravity, vol. 2, p. 611. Clarendon Press, Oxford (1981). Reprinted in [42, p 117]
15. Bell, J.S.: De Broglie-Bohm, delayed choice double slit experiment and density matrix. Int. J. Quantum Chem. Quantum Chemistry Symposium **14**, 155–159 (1980). Reprinted in [42, p. 111]
16. Bell, J.S.: In: M. Flato et al. (eds) Quantum mechanics, determinism, causality, and particles. D. Reidel, Dordrecht-Holland, 11–17 (1976). Reprinted in [42, p. 93].
17. Bohm, D.: A suggested interpretation of the quantum theory in terms of "hidden" variables. Phys. Rev. **85**, 166, 180 (1952)
18. Bohm, D.: Quantum Theory. Prentice Hall, Englewood Cliffs (1951)
19. Wheeler, J.A., Zurek,W.H. (eds.): Quantum Theory and Measurement. Princeton University Press, Princeton (1983)
20. Jarrett, J.: On the physical significance of the locality conditions in the Bell arguments. Nous **18**, 569–589 (1984)
21. Evans, P., Price, H., Wharton, K.B.: New slant on the EPR-Bell experiment. Physics. arXiv:1001.5057v3 [quant-ph].

22. Bethe, H.: My experience in teaching physics. Am. J. Phys. **61**, 972 (1993)
23. Gell-Mann, M.: The Quark and the Jaguar: Adventures in the Simple and the Complex. W.H. Freeman, New York (1994)
24. Messiah, A.: Quantum Mechanics, Volumes I and II. North-Holland Publishing Company, Amsterdam. English translation by G.M. Temmer published by John Wiley and Sons, New York, 1976 (1959)
25. Shankar, R.: Principles of Quantum Mechanics, 2nd edn. Plenum Press, New York (1994)
26. Gilder, L.: The Age of Entanglement: When Quantum Physics Was Reborn. Alfred A. Knopf, New York (2008)
27. Freedman, S.J., Clauser, J.F.: Experimental test of local hidden-variable theories. Phys. Rev. Lett. **28**, 938 (1972)
28. Fry, E.S., Thompson, R.C.: Experimental test of local hidden-variable theories. Phys. Rev. Lett. **37**, 465 (1976)
29. Aspect, A., Grangier, P., Roger, G.: Experimental tests of realistic local theories via Bell's theorem. Phys. Rev. Lett. **47**, 460 (1981)
30. Aspect, A., Grangier, P., Roger, G.: Experimental realization of Einstein-Podolsky-Rosen gedankenexperiment: a new violation of Bell's inequalities. Phys. Rev. Lett. **49**, 91 (1982)
31. Aspect, A., Dalibard, J., Roger, G.: Experimental test of Bell's inequalities using time-varying analyzers. Phys. Rev. Lett. **49**, 1804 (1982)
32. Clauser, J.F., Horne, M., Holt, R.A., Shimony, A.: Proposed Experiment to test local hidden-variable theories. Phys. Rev. Lett. **23**, 880 (1969)
33. Lamehi-Rachti, M., Mittig, W.: Quantum mechanics and hidden variables: a test of Bell's inequality by the measurement of the spin correlation in low-energy proton-proton scattering. Phys. Rev. **D**, 2543–2555 (1976)
34. Bell, J.S.: Einstein-Podolsky-Rosen experiments. In: Proceedings on the Frontier Problems in High Energy Physics, Pisa, pp. 33–45 (1976). Reprinted in [41, p. 768]
35. Herbert, N.: Quantum Reality. Doubleday, New York (1985)
36. Davies, P.C.W., Brown, J.R.: The Ghost in the Atom. Cambridge University Press, Cambridge (1986)
37. Bell, J.S.: Free variables and local causality. Epistemol. Lett. **15** (1977)
38. Bell, J.S.: Free variables and local causality. Dialectica **39**, 103–106 (1985). This paper is also presented in [41, p. 778–781]
39. Goldstein, S., Norsen, T., Tausk, D., Zanghi, N.: "Bell's Theorem" in Scholarpedia peer-reviewed online encyclopedia.www.scholarpedia.org/article/Bell_theorem
40. Ghirardi, G.C.: Sneaking a Look at God's Cards. Princeton University Press, Princeton (2005)
41. Bell, M., Gottfried, K., Veltman, M. (eds.): Quantum Mechanics High Energy Physics and Accelerators: Selected Papers of John S. Bell (with commentary). World Scientifc Publishing, Singapore (1995)
42. Bell, J.S.: Speakable and Unspeakable in Quantum Mechanics. Cambridge University Press, Cambridge (1987). Many of the works by Bell which are of concern to us may be found in this reference. See also [41] and [43]. The latter two are complete collections containing all of Bell's papers on quantum foundations
43. Bell, M. Gottfried, K., Veltman, M. (eds.): John S. Bell on the Foundations of Quantum Mechanics. World Scientifc Publishing, Singapore (2001)

Chapter 4
Schrödinger's Paradox and Nonlocality

4.1 Introduction

In the previous chapter, we reviewed the Einstein–Podolsky–Rosen paradox and Bell's Theorem. EPR is a very powerful result, and when its logical content is distilled, this famous analysis leads to a striking conclusion—the existence of physical properties lying outside the purview of the quantum description. Bell's theorem essentially embodies a proof that such physical properties cannot be made to agree with quantum mechanics, but must lead to predictions that contradict it. Upon examining the two arguments, we saw that they imply quantum nonlocality.

We now proceed[1] with our exploration of this topic by pursuing the direction first investigated by Erwin Schrödinger in 1935 and 1936. In a series of papers [8–10] published just after the Einstein–Podolsky–Rosen paradox, Schrödinger was able to extend their result in a very significant way, demonstrating that the quantum state these authors examined displays perfect correlations[2] between *all* observables associated with each particle. Just as we did in the case of the Einstein–Podolsky–Rosen paradox, we will find Schrödinger's result leads to an incompleteness argument according to which there must exist elements of reality on the observables. We shall call this result the "*Schrödinger paradox*," for parallelism with the usual appellation given the famous analysis of Einstein, Podolsky and Rosen.

Further, we show that the type of correlations found by Schrödinger are also displayed by any one of the so-called "maximally entangled states" and we analyze this assertion for a general case. From this rich variety of results springs the possibility

---

[1] Among the original results presented in this chapter are the nonlocality proofs to be presented in Sect. 4.5. Similar results were discovered by Brown and Svetlichny [1], Heywood and Redhead [2], Aravind [3], and by Cabello [4]. Our analysis of the Conway Kochen Free Will Theorem in Sect. 4.5.2 is related to the criticism given by Bassi and Ghirardi [5], Tumulka [6] and also Goldstein et al. [7].

[2] The quantum state examined by Einstein, Podolsky and Rosen is not the spin singlet state, but a state which shows correspondence between measurements of position and momenta of two particles.

D. L. Hemmick and A. M. Shakur, *Bell's Theorem and Quantum Realism*,
SpringerBriefs in Physics, DOI: 10.1007/978-3-642-23468-2_4,
© The Author(s) 2012

for a large class of new nonlocality proofs, as well. When we reach the stage at which such issues have been mastered, this will also grant us a very penetrating vision of the Conway and Kochen Free Will Theorem. It will then be straightforward to show that, while the theorem contains no mathematical flaws, the claims made by its authors regarding quantum physics are not supported.

In Chap. 3, we did not approach the Einstein–Podolsky–Rosen paradox via the quantum state they addressed, but we considered instead the spin-singlet version of the famous analysis (due to Bohm [11]). Therefore, we shall begin this chapter with a presentation of the quantum system originally analyzed by EPR. We then show how one may develop the results first noted by Schrödinger, i.e., that the state exhibits additional perfect correlations beyond those noted by Einstein, Podolsky and Rosen. We then present the incompleteness argument that follows from such perfect correlations. Next in our program, we proceed to generalize the Schrödinger paradox by developing different forms of maximally entangled states and also demonstrate how the perfectly correlated observables arise. At this point, we discuss Schrödinger nonlocality proofs and the Conway and Kochen Free Will Theorem.

The chapter ends with the presentation of yet another of Schrödinger's intriguing discussions of hidden variables from his "cat paradox" paper [8]. Interestingly, this particular discourse bears a close resemblance to von Neumann's no-hidden-variables theorem, although Schrödinger does not repeat von Neumann's error. Nevertheless, the conclusions at which he does arrive are quite curious, and evoke further speculations.

4.2 Schrödinger's Generalization of EPR

4.2.1 The Einstein–Podolsky–Rosen Quantum State

The Bohm version of the EPR paradox presented in Chap. 3 was addressed to the spin components of two particles represented by the spin singlet state. While this argument was concerned with perfect correlations in the spin components, the original form of the EPR paradox involved perfect correlations of a slightly different form for the positions and momenta. For a system described by the original EPR state, we find that measurements of the positions x_1 and x_2 of the two particles give equal results,[3,4] while measurements of the momenta give results which sum to zero.

---

[3]   EPR discuss a form in which the difference between the positions is equal to a *constant* they call d. The case we consider differs from this only in that the points of origin from which the positions of the two particles are measured are different, so that the quantity $x_2 - x_1 + d$ for EPR becomes $x_2 - x_1$ for our case.

[4]   It is clear that the Einstein–Podolsky–Rosen quantum state (4.3) is not normalizable. Nevertheless, as we saw in the example of the spin singlet state, it is possible to carry out a similar argument for states which *can* be normalized. The 'maximally entangled states', which are the subject of the Schrödinger paradox, contain among them a large class of normalizable states, as we shall see.

To develop the perfect correlations in position and momentum, we first recall how the spin singlet state leads to the spin correlations. The spin singlet state takes the form (3.6):

$$\psi_{ss} = |\uparrow\theta,\phi\rangle \otimes |\downarrow\theta,\phi\rangle - |\downarrow\theta,\phi\rangle \otimes |\uparrow\theta,\phi\rangle, \qquad (4.1)$$

where we have suppressed the normalization for simplicity. We have here a sum of two terms, each being a product of an eigenfunction of $\sigma_{\theta,\phi}^{(1)}$ with an eigenfunction of $\sigma_{\theta,\phi}^{(2)}$ such that the corresponding eigenvalues sum to zero. In the first term of (4.1) for example, the factors are eigenvectors corresponding to $\sigma_{\theta,\phi}^{(1)} = \frac{1}{2}$ and $\sigma_{\theta,\phi}^{(2)} = -\frac{1}{2}$. With this, it is clear that we will have perfect correlations between the results of measurement of $\sigma_{\theta,\phi}^{(1)}$ and $\sigma_{\theta,\phi}^{(2)}$, i.e. measurements of these observables will give results which sum to zero. By analogy with this result, the perfect correlations in position and momentum emphasized by Einstein, Podolsky, and Rosen will follow if the wave function assumes the forms:

$$\psi = \int_{-\infty}^{\infty} dp \, |\phi_{-p}\rangle \otimes |\phi_p\rangle$$

$$\psi = \int_{-\infty}^{\infty} dx \, |\varphi_x\rangle \otimes |\varphi_x\rangle. \qquad (4.2)$$

Here $|\phi_p\rangle$ is the eigenvector of momentum operator corresponding to a momentum of p. $|\varphi_x\rangle$ is the eigenvector of position. We now show that the wave function given by the first equation in (4.2):

$$\psi_{EPR} = \frac{1}{2\pi\hbar} \int_{-\infty}^{\infty} dp |\phi_{-p}\rangle \otimes |\phi_p\rangle \qquad (4.3)$$

assumes also the form of the second equation in (4.2).

To see this we expand the first factor in the summand of ψ_{EPR} in terms of $|\varphi_x\rangle$:

$$\psi_{EPR} = \int_{-\infty}^{\infty} dp \left(\int_{-\infty}^{\infty} dx |\varphi_x\rangle\langle\varphi_x|\phi_{-p}\rangle \right) \otimes |\phi_p\rangle. \qquad (4.4)$$

Using $\langle\varphi_x|\phi_{-p}\rangle = \langle\phi_p|\varphi_x\rangle$, this becomes

$$\psi_{EPR} = \int_{-\infty}^{\infty} dx |\varphi_x\rangle \otimes \int_{-\infty}^{\infty} dp |\phi_p\rangle\langle\phi_p|\varphi_x\rangle. \qquad (4.5)$$

Since $\int_{-\infty}^{\infty} dp|\phi_p\rangle\langle\phi_p|$ is a unit operator, it follows that

$$\psi_{EPR} = \int\limits_{-\infty}^{\infty} dx|\varphi_x\rangle \otimes |\varphi_x\rangle \qquad (4.6)$$

This is the result we had set out to obtain.

4.2.2 Schrödinger's Generalization and Maximal Perfect Correlations

Schrödinger's work essentially revealed the full potential of the quantum state which Einstein, Podolsky, and Rosen had considered. In his analysis, Schrödinger demonstrated that the perfect correlations the EPR state exhibits are not limited to those in the positions and momenta. For two particles described by the EPR state, *every* observable of each particle will exhibit perfect correlations with an observable of the other. What we present here and in subsequent sections is a simpler way to develop such perfect correlations than that given by Schrödinger[5] This result may be seen as follows. The EPR state (4.3) is rewritten as[6]

$$\langle x_1, x_2|\psi_{EPR}\rangle = \int\limits_{-\infty}^{\infty} dp_1 dp_2\langle x_1, x_2|p_1, p_2\rangle\langle p_1, p_2|\psi_{EPR}\rangle = \delta(x_2 - x_1). \quad (4.7)$$

We may then use a relationship known as the *completeness* relationship. According to the completeness relation, if $\{\phi_n(x)\}$ is any basis for the Hilbert space L_2, then we have $\sum_{n=1}^{\infty} \phi_n^*(x_1)\phi_n(x_2) = \delta(x_2 - x_1)$. The EPR state may be rewritten using this relation as:

$$\psi_{EPR}(x_1, x_2) = \sum_{n=1}^{\infty} \phi_n^*(x_1)\phi_n(x_2), \qquad (4.8)$$

where $\{\phi_n(x)\}$ is an *arbitrary* basis of L_2. Since the form (4.8) resembles the spin singlet form (3.6), one might anticipate that it will lead to the existence of perfect correlations between the observables of particles 1 and 2. As we shall see, this is true not only for quantum systems described by (4.8), but also for a more general class of states as well.

---

[5] See [9].

[6] The reader may object that for this state, the two particles lie 'on top of one another', i.e., the probability that $x_1 \neq x_2$ is identically 0. However, in the following sections, we show that the same conclusions drawn for the EPR state also follow for a more general class of 'maximally entangled states', of which many do *not* restrict the positions in just such a way.

Let us consider a Hermitian operator A (assumed to possess a discrete spectrum) on the Hilbert space L_2. At this point, we postpone defining an observable of the EPR system in terms of A—we simply regard A as an abstract operator. Suppose that A can be written as

$$A = \sum_{n=1}^{\infty} \mu_n |\phi_n\rangle \langle \phi_n|, \tag{4.9}$$

where $|\phi_n\rangle \langle \phi_n|$ is the one-dimensional projection operator associated with the vector $|\phi_n\rangle$. The eigenvectors and eigenvalues of A are, respectively, the sets $\{|\phi_n\rangle\}$, and $\{\mu_n\}$. Suppose that another Hermitian operator, called $\tilde{A}$, is defined by the relationship

$$\tilde{A}_{x,x'} = A^*_{x,x'}, \tag{4.10}$$

where $\tilde{A}_{x,x'}$, and $A_{x,x'}$ are, respectively, the matrix elements of $\tilde{A}$ and A in the position basis $\{\varphi_x\}$, and the superscript '$*$' denotes complex conjugation. Note that for any given A, the 'complex conjugate' operator $\tilde{A}$ defined by (4.10) is *unique*. Writing out $A_{x,x'}$ gives

$$A_{x,x'} = \langle \varphi_x | \left(\sum_{n=1}^{\infty} \mu_n |\phi_n\rangle\langle \phi_n| \right) |\varphi_{x'}\rangle$$

$$= \sum_{n=1}^{\infty} \mu_n \langle \varphi_x|\phi_n\rangle \langle \phi_n|\varphi_{x'}\rangle. \tag{4.11}$$

Using (4.10), we find

$$\tilde{A}_{x,x'} = \sum_{n=1}^{\infty} \mu_n \langle \varphi_x|\phi_n\rangle^* \langle \phi_n|\varphi_{x'}\rangle^*$$

$$= \sum_{n=1}^{\infty} \mu_n \langle \varphi_x|\phi_n^*\rangle \langle \phi_n^*|\varphi_{x'}\rangle, \tag{4.12}$$

where $|\phi_n^*\rangle$ is just the Hilbert space vector corresponding[7] to the function $\phi_n^*(x)$. From the second equation in (4.12), it follows that $\tilde{A}$ takes the form

---

[7] More formally, we develop the correspondence of Hilbert space vectors $|\phi\rangle$ to functions $\phi(x)$ by expanding $|\psi\rangle$ in terms of the position eigenvectors $|\varphi_x\rangle$:

$$|\phi\rangle = \int_{-\infty}^{\infty} dx |\varphi_x\rangle\langle \varphi_x|\phi\rangle. \tag{4.13}$$

$$\tilde{A} = \sum_{n=1}^{\infty} \mu_n |\phi_n^*\rangle\langle\phi_n^*|. \qquad (4.15)$$

The eigenvectors and eigenvalues of $\tilde{A}$ are, respectively, the sets $\{|\phi_n^*\rangle\}$, and $\{\mu_n\}$.

We now consider the observables $\mathbf{1} \otimes A$ and $\tilde{A} \otimes \mathbf{1}$ on the Hilbert space of the EPR state, $L_2 \otimes L_2$, where $\mathbf{1}$ is the identity operator on L_2. Put less formally, we consider A to be an observable of particle-2, and $\tilde{A}$ as an observable of particle-1, just as $\sigma_{\hat{a}}^{(1)}$ represented 'the spin component of particle-1' in our above discussion of the spin singlet state. Examining (4.8), we see that each term is a product of an eigenvector of A with an eigenvector of $\tilde{A}$ such that the eigenvalues are equal. For these observables, we have that ψ_{EPR} *is an eigenstate of $A - \tilde{A}$ of eigenvalue zero*, i.e., $(A - \tilde{A})\psi_{EPR} = 0$. Thus, ψ_{EPR} exhibits *perfect correlation* between A and $\tilde{A}$ in that the measurements of A and $\tilde{A}$ give results that are equal.

Recall that the L_2 basis $\{\phi_n(x)\}$ appearing in (4.8) is an arbitrary L_2 basis. With this, and the fact that the eigenvalues μ_n chosen for A are arbitrary, it follows that the operator A can represent *any* observable of particle-2. We can thus conclude that for any observable A of particle-2, there exists a unique observable $\tilde{A}$ (defined by (4.10)) of particle-1 which exhibits perfect correlations with the former.

We may interchange the roles of particles 1 and 2 in the above argument, if we note that the EPR state is *symmetric* in x_1 and x_2. This latter is proved as follows. Since the Dirac delta function is an even function, we have

$$\delta(x_2 - x_1) = \delta(x_1 - x_2). \qquad (4.16)$$

Thus the EPR state assumes the form

$$\psi_{EPR} = \delta(x_2 - x_1) = \delta(x_1 - x_2) = \sum_{n=1}^{\infty} \phi_n(x_1)\phi_n^*(x_2), \qquad (4.17)$$

where $\{\phi_n(x)\}$ is an arbitrary basis of L_2, and the second equality is the completeness relation. This establishes the desired symmetry.

Suppose now that we consider the observables $A \otimes \mathbf{1}$ and $\mathbf{1} \otimes \tilde{A}$, i.e., we consider A as an observable of particle-1, and $\tilde{A}$ as an observable of particle-2. Then from

---

Footnote 7 (Continued)
The function $\phi(x)$ can then be identified with the inner product $\langle\varphi_x|\phi\rangle$. Then the Hilbert space vector corresponding to $\phi^*(x)$ is just

$$\begin{aligned}
|\phi^*\rangle &= \int_{-\infty}^{\infty} dx |\varphi_x\rangle\langle\varphi_x|\phi\rangle^* \\
&= \int_{-\infty}^{\infty} dx |\varphi_x\rangle\langle\phi|\varphi_x\rangle.
\end{aligned} \qquad (4.14)$$

the symmetry of the EPR state, it follows that we can reverse the roles of the particles in the above discussion to show that for any observable A of particle-1, there exists a unique observable $\tilde{A}$ of particle-2 defined by (4.10), which exhibits perfect correlations with the former.

The perfect correlations in position and momentum originally noted by Einstein, Podolsky and Rosen are a special case of the perfectly correlated observables A and $\tilde{A}$ given here. EPR found that the measurement of the positions x_1 and x_2 of the two particles must give results that are *equal*. The momenta p_1 and p_2 showed slightly different perfect correlations in which their measurements give results which sum to zero. To assess whether the EPR perfect correlations in position and momentum are consistent with the scheme given above, we derive using (4.10) the forms of $\tilde{x}$ and $\tilde{p}$. Since the position observable x is diagonal in the position eigenvectors and has real matrix elements, (4.10) implies that $\tilde{x} = x$. As for the momentum, this operator is represented in the position basis by the differential operator $-i\hbar\frac{d}{dx}$. Using (4.10) we see that $\tilde{p}$ is just equal to the *negative* of p itself. Hence the perfect correlations developed in the present section, according to which A and $\tilde{A}$ are equal, imply that measurements of the positions of the two particles must give equal results, whereas measurements of the momenta p_1 and p_2 must give results which sum to zero. This is just what EPR had found.

4.3 Schrödinger's Paradox, and Incompleteness

The existence of such ubiquitous perfect correlations allows one to develop an argument demonstrating that all observables of particle-1 and all observables of particle-2 are "elements of reality", i.e., they possess definite values.[8] This argument is similar to the incompleteness argument given above for the spin singlet EPR case.

Consider the possibility of separate measurements of observables $\tilde{A}$, A being performed, respectively, on particles 1 and 2 of the EPR state. As in the spin singlet EPR analysis, we assume locality, so that the properties of each particle must be regarded as independent of those of its spatially separated partner. Imagine that observable $\tilde{A}$ of particle-1 is measured. Because $\tilde{A}$ is perfectly correlated with A of particle-2, the result we find allows us to predict with certainty the result of any subsequent measurement of the latter. If, for example, measurement of $\tilde{A}$ gives the result $\tilde{A} = \mu_a$ then we can predict with certainty that measuring A will give $A = \mu_a$. Since we have assumed *locality*, A must be an element of reality, i.e., there must be some definite value associated with it.

As in the discussion of the EPR paradox in Chap. 3 (Sect. 3.2.2) we emphasize once again that this form of argument is rich with meaning and vital to the subsequent

---

[8] In the previous chapter, we noted that one must guard against taking the EPR incompleteness as "the final word." The true state of affairs in the matter may be seen when EPR is considered *in conjunction with Bell's theorem*. From such a complex argument, it follows that quantum mechanics is irreducibly nonlocal. Similarly, the Schrödinger incompleteness argument must be taken as the first part of another complex argument leading to nonlocality. This is to be discussed in Sect. 4.5.

analyses we shall offer regarding quantum nonlocality. It is difficult to overstate the importance of these results.

To forecast the value of A with perfect certainty means that one may test ensembles of such states, performing arbitrarily many trials and the prediction will never fail. It is clear that some matter of fact obtains regarding the value of the observable.

Such a predetermined value is quite crucial, existing beyond the purview of quantum mechanics, and leading to novel physical and mathematical implications. The quantum mechanical state description, as prescribed by ψ, does not allow for the definite status of an observable prior to measurement.[9] Therefore, to account for the physical situation, it becomes necessary to introduce a new mathematical representation. To conform with notation used previously, we let λ supplement the mathematics, so that the new description of state consists of both ψ and λ.

The existence of a predetermined value for each observable A of particle-2 is equivalent to the existence of what we have been calling a value-map function $V_\lambda(A)$.[10] For each fixed λ, we have an assignment $V(O)$ from the observables to values, which we rewrite as $E(O)$ for consistency with the notation of Chaps. 1 and 2.

Furthermore, as we saw in the previous section, the perfect correlations are also universal in terms of the observables of particle-1. One may select an arbitrary observable A of that particle and demonstrate the structure of the quantum state given in Eq. 4.8 implies that there is a unique observable $\tilde{A}$ of particle-2 with which A is perfectly correlated.

That being the case, one may carry out arguments analogous to those given in this section in which the roles of particle-1 and particle-2 are *reversed*. In this way one can demonstrate that a value map $E(O)$ must also exist for all observables of particle-1.[11]

4.3.1 Perfect Correlations and Procedure of Measurement

In the discussions of this chapter, one important aspect of the quantum formalism has been given little attention. Being as compact and straightforward as it is, the Schrödinger paradox analysis might easily allow this facet of things to

---

[9] Some might dissent, saying that if ψ happens to be an eigenstate of an observable then quantum theory does make such an allowance. However, such an assertion carries very little weight. Firstly, while its true that eigenstates are special in that they lead to precise results in each case, the formal statement of quantum theory assigns no meaning to a physical quantity apart from results of measurement. Secondly, even if one were still to insist that eigenstates are somehow special exceptions to what quantum theory tells us, the system in question at present is not described by such a state.

[10] As in the EPR case, V has no dependence upon ψ since we are again considering a fixed wave-function, namely the original EPR state.

[11] We have made these considerations based on analysis of the EPR quantum state. Yet, one may develop maximally entangled states in a great variety of ways, as is shown in the section to follow. In particular, there is no need to limit the argument to infinite-dimensional cases, as we shall see. For every maximally entangled state, the Schrödinger paradox argument may be carried out and the same conclusions will apply.

escape our notice. While the quantum formalism's definition of physical quantities is straightforward—each self adjoint operator corresponds to an observable—one finds that the rules of measurement contain something of a subtlety. As we saw in Sect. 1.3, for each quantum observable, the formalism often allows for a variety of *distinct* measurement procedures.

For the sake of discussion, it will be helpful to have a specific notation to refer to a measurement procedure. For this purpose, we call $\mathcal{M}(A)$ a measurement procedure used in determining the value of A in a quantum experiment. The perfect correlations so far worked out have concerned an arbitrary observable A of particle-2 and the perfectly correlated observable $\tilde{A}$ of particle-1. The question we would like to examine is the following. Does the issue of measurement procedure play any role in the perfect correlations which Schrödinger demonstrated?

In fact, it is not difficult to see that measurement procedure plays no role whatever. The perfect correlations follow very easily from the structure of the quantum state as displayed in Eq. 4.8. As was mentioned in the discussion given in that section, the state is an eigenstate of the difference of the two observables, and in particular we have $(A - \tilde{A})\psi = 0$. In other words, the fact that measurements of A and $\tilde{A}$ are always equal is *independent* of what choice of measurement procedure one might select for either A or for $\tilde{A}$.

This being the case, it follows that the value assignment map $E(O)$ is non-contextual. Hence, the conclusion of the Schrödinger paradox is just the same as the hidden variables formulation appearing in the theorems of Gleason, of Kochen and Specker, and indeed of *all* of the spectral incompatibility theorems. The implications of this fact will be drawn out in subsequent sections.

The key to the above incompleteness argument is that one can, by measurement of an observable of particle-2, predict with certainty the result of any measurement of a particular observable of particle-1. In his presentation of the EPR incompleteness argument (which concerns position and momentum) Schrödinger uses a colorful analogy to make the situation clear. He imagines particles 1 and 2 as being a student and his instructor. The measurement of particle-2 corresponds to the instructor consulting a textbook to check the answer to an examination question, and the measurement of particle-1 to the response the student gives to this question. Since he always gives the correct answer, i.e., the same as that in the instructor's textbook, the student must have known the answer beforehand. Schrödinger presents the situation as follows: [8] (emphasis by original author).

> Let us focus attention on the system labeled with small letters p,q and call it for brevity the "small" system. Then things stand as follows. I can direct *one* of two questions to the small system, either that about q or that about p. Before doing so I can, if I choose, procure the answer to *one* of these questions by a measurement on the fully separated other system (which we shall regard as auxiliary apparatus), or I may take care of this afterwards. My small system, like a schoolboy under examination, *cannot possibly know* whether I have done this or for which questions, or whether or for which I intend to do it later. From arbitrarily many pretrials I know that the pupil will correctly answer the first question I put to him. From that it follows that in every case, he *knows* the answer to *both* questions. ... No school principal would judge otherwise ... He would not come to think that his, the teacher's, consulting a textbook first suggests to the pupil the correct answer, or even, in the cases when the teacher

chooses to consult it only after ensuing answers from the pupil, that the pupil's answer has changed the text of the notebook in the pupil's favor.

4.3.2 Schrödinger's Theorem

As in the case of the Einstein–Podolsky–Rosen paradox (see Sect. 3.2.3), Schrödinger's generalization and extension of the argument is also equivalent to a logical theorem,[12] and we now develop this explicitly.

The structure of the Schrödinger paradox is essentially parallel to that of the Einstein–Podolsky–Rosen paradox. The notion of an "element of reality" once again plays a central role in the analysis. Because of its importance, it is worth repeating here the EPR sufficient condition [12] "If without in any way disturbing a system, we can predict with certainty (i.e., with probability equal to unity) the value of a physical quantity, then there exists an element of physical reality corresponding to this quantity."

Note that this condition entails two criteria to be met: that prediction is made in a passive way, i.e., with no disturbance, and that prediction be made with perfect certainty. We have discussed the physical state of affairs in terms of a two-particle system such that separate measurements are to be carried out on each. We imagined that spatial separation was enough to guarantee each such process may be considered independent of any immediate influence from the other. This autonomy is called on as we assert that measurement of one particle does not "in any way disturb" the second.

It is easy to see that the linking of spatial separation with freedom from influence depends upon locality. To drop the notion of locality[13] would be to permit mutual instantaneous influences between the particles, which would destroy their claim to independence.

---

[12] We present the theorem below and assert that its assumptions are locality and the perfect correlations. However, one can point to two additional background assumptions which also lie behind the theorem. First, as mentioned in Chap. 3, Sect. 3.4.3, there is the 'no-superdeterminism' assumption. Secondly, there is the matter of which elements of quantum mechanics are required to carry through the argument. We will discover that it is not necessary to carry the full apparatus of quantum mechanics as a background assumption for Schrödinger paradox and Schrödinger's Theorem. All that is needed is the quantum rule stipulating that results of procedures measuring observables or sets of compatible observables are the appropriate eigenvalues or joint-eigenvalues. This will be important in Sect. 4.5.2.

[13] For those concerned with minimizing the assumptions, we might point out that the argument here may be made using a weaker version of locality. Locality, as one ordinarily defines it, requires that no information of any kind may be transmitted faster than light speed. What is needed for Schrödinger's Theorem, is simply that certain types of disturbances cannot influence distant events. Suppose that the observable A is measured by the procedure $\mathcal{M}(A)$, while the observable $\tilde{A}$ of the second particle is to be measured. In this scenario, we suppose that the two systems are *sufficiently independent* such that the choice of measurement procedure $\mathcal{M}$ cannot influence the result of the measurement of $\tilde{A}$. Such modified or limited locality might be termed "measurement locality," for brevity.

"Prediction with certainty" is the second criterion for an element of reality. In fact, the perfect correlations are sufficient to guarantee such a demand: every observable A of either particle has a partner $\tilde{A}$ associated with the other particle such that measurements of the two yield equal results. As we have discussed, this equality cannot be affected in any way by the choice of measurement procedure. It might be appropriate to call these *context-independent* perfect correlations.

Since both criteria have been met, it follows that all the quantities subject to such perfect correlation must be elements of reality. In the case of the Schrödinger paradox, we have that for all physical quantities associated with each particle, there must exist explicit values which preestablish the measurement results.

To sum up, the logical components of the Schrödinger paradox are as follows. Firstly, the procedures involved in the predictions must not disturb the system. Since the particles are spatially separated, this criterion is guaranteed by locality. Secondly, we must have that a physical quantity may be predicted with certainty, and this is given to us by the (noncontextual) perfect correlations on the maximally entangled EPR quantum state. Based on these conditions, it follows that there exist noncontextual hidden variables on all physical quantities of each of the two particles.

Just as the Einstein–Podolsky–Rosen paradox revealed itself equivalent to a theorem, we have found that the same situation is true for the Schrödinger paradox. The term "Schrödinger's Theorem" seems quite appropriate.

Having developed this result, we are led to reflect on the power of Schrödinger's analysis. The assumptions of Schrödinger's Theorem are not overly stringent, and one might imagine a large family of potential theories which meet them. This will be important when we come to Schrödinger nonlocality in Sect. 4.5.

4.4 EPR Quantum State and Other Maximally Entangled States

4.4.1 Generalized Form of the EPR State

Consideration of the structure of the EPR state (4.8) suggests the possibility that a more general class of states might also exhibit maximal perfect correlations. Before developing this, we first note that the formal way to write (4.8) is such that each term consists of a tensor product of a vector of particle-1's Hilbert space with a vector of particle-2's Hilbert space:

$$\psi_{EPR} = \sum_{n=1}^{\infty} |\phi_n^*\rangle \otimes |\phi_n\rangle. \tag{4.18}$$

We shall often use this convenient notation in expressing the maximally entangled states. The operation of complex conjugation in the first factor of the tensor product may be regarded as a special case of a class of operators known as "anti-unitary involutions"—operators which we shall denote by C, to suggest complex conjugation. We will find that any state of the form

$$\psi_C = \sum_{n=1}^{\infty} C|\phi_n\rangle \otimes |\phi_n\rangle, \qquad (4.19)$$

will exhibit ubiquitous perfect correlations leading to the conclusion of definite values on all observables of both subsystems.

An *anti-unitary involution* operation represents the generalization of complex conjugation from scalars to vectors. The term "involution" refers to any operator C whose square is equal to the identity operator, i.e. $C^2 = 1$. Anti-unitarity entails two conditions, the first of which is anti-linearity:

$$C(c_1|\psi_1\rangle + c_2|\psi_2\rangle + \cdots) = c_1^* C|\psi_1\rangle + c_2^* C|\psi_2\rangle + \cdots, \qquad (4.20)$$

where $\{c_i\}$ are constants and $\{|\psi_i\rangle\}$ are vectors. The second condition is the anti-linear counterpart of unitarity:

$$\langle C\psi|C\phi\rangle = \langle \psi|\phi\rangle^* = \langle \phi|\psi\rangle \quad \forall \psi, \phi, \qquad (4.21)$$

which tells us that under the operation of C, inner products are replaced by their complex conjugates. Note that this property is sufficient to guarantee that if the set $\{|\phi_n\rangle\}$ is a basis, then the vectors $\{C|\phi_n\rangle\}$ in (4.19) must also form a basis. For each anti-unitary involution C, there is a special Hilbert space basis whose elements are invariant under C. The operation of C on any given vector $|\psi\rangle$ can be easily obtained by expanding the vector in terms of this basis. If $\{\varphi_n\}$ is this special basis then we have

$$C|\psi\rangle = C \sum_{i=1}^{\infty} |\varphi_i\rangle\langle\varphi_i|\psi\rangle = \sum_{i=1}^{\infty} |\varphi_i\rangle\langle\varphi_i|\psi\rangle^*. \qquad (4.22)$$

When one is analyzing any given state of the form (4.19), it is convenient to express the state and observables using this special basis. The EPR state is a special case of the state (4.19) in which the anti-unitary involution C is such that the position basis $\{|\varphi_x\rangle\}$ plays this role.

The state (4.19) shows an invariance similar to that we developed for the EPR state: the basis $\{|\phi_n\rangle\}$ in terms of which the state is expressed, is *arbitrary*. We now develop this result. Note that the expression (4.19) takes the form

$$\psi_C = \sum_{n=1}^{\infty} C \left(\sum_{i=1}^{\infty} |\chi_i\rangle\langle\chi_i|\phi_n\rangle \right) \otimes \left(\sum_{j=1}^{\infty} |\chi_j\rangle\langle\chi_j|\phi_n\rangle \right), \qquad (4.23)$$

if we expand the vectors in terms of an arbitrary basis $\{\chi_i\}$. Applying the C operation in the first factor and rearranging the expression, we obtain

$$\psi_C = \sum_{n=1}^{\infty}\left(\sum_{i=1}^{\infty} C|\chi_i\rangle\langle\phi_n|\chi_i\rangle\right) \otimes \left(\sum_{j=1}^{\infty} |\chi_j\rangle\langle\chi_j|\phi_n\rangle\right)$$

$$= \sum_{i=1}^{\infty}\sum_{j=1}^{\infty}\sum_{n=1}^{\infty}\langle\chi_i|\phi_n\rangle\langle\phi_n|\chi_j\rangle C|\chi_i\rangle \otimes |\chi_j\rangle$$

$$= \sum_{i=1}^{\infty}\sum_{j=1}^{\infty}\langle\chi_i|\left(\sum_{n=1}^{\infty}|\phi_n\rangle\langle\phi_n|\right)|\chi_j\rangle C|\chi_i\rangle \otimes |\chi_j\rangle,$$

$$(4.24)$$

where the first equality follows from the anti-linearity of C. Since the expression $\sum_{n=1}^{\infty}|\phi_n\rangle\langle\phi_n|$ is the identity operator, the orthonormality of the set $\{\chi_i\}$ implies that

$$\psi_C = \sum_{i=1}^{\infty}\sum_{j=1}^{\infty}\delta_{ij}C|\chi_i\rangle \otimes |\chi_j\rangle = \sum_{i=1}^{\infty}C|\chi_i\rangle \otimes |\chi_i\rangle. \tag{4.25}$$

Thus, the form of the state (4.19) is invariant under any change of basis, and we have the desired result.

Note then the role played by the properties of anti-linear unitarity and anti-linearity: from the former followed the result that the vectors $\{C|\phi_n\rangle\}$ form a basis if the vectors $\{|\phi_n\rangle\}$ do so, and from the latter followed the invariance just shown.

Consider a Hermitian operator A on L_2 which can be written as

$$A = \sum_{n=1}^{\infty}\mu_n|\phi_n\rangle\langle\phi_n|. \tag{4.26}$$

Note that A's eigenvectors and eigenvalues are given by the sets $\{|\phi_n\rangle\}$ and $\{\mu_n\}$, respectively. If we define the observable $\tilde{A}$ by the relationship

$$\tilde{A} = CAC^{-1}, \tag{4.27}$$

then $\tilde{A}$'s eigenvalues are the same as A's, i.e., $\{\mu_n\}$, and its eigenvectors are given by $\{C|\phi_n\rangle\}$. To see this, note that

$$CAC^{-1}C|\phi_n\rangle = C\mu_n|\phi_n\rangle = \mu_n C|\phi_n\rangle, \tag{4.28}$$

where the first equality follows from $CC^{-1} = 1$ and $A|\phi_n\rangle = \mu_n\phi_n$. Since $C^2 = 1$, it follows that $C = C^{-1}$, and we may rewrite (4.27) as[14]:

$$\tilde{A} = CAC. \tag{4.31}$$

---

[14] Note that to evaluate $\tilde{A}$, one can express it using its matrix elements with respect to the invariant basis $\{\varphi_n\}$ of C. If we evaluate the matrix element CAC_{ij}, we find

$$\langle\varphi_i|CAC\varphi_j\rangle = \langle\varphi_i|CA\varphi_j\rangle = \langle\varphi_i|A\varphi_j\rangle^*, \tag{4.29}$$

Note that for any given A, the observable $\tilde{A}$ defined by (4.31) is unique.

If we identify A as an observable of subsystem 2, and $\tilde{A}$ as an observable of subsystem 1, then examination of the state (4.19) shows that these exhibit *perfect correlations* such that their measurements always yield results that are equal. From the invariance of the state, we can then conclude that for *any* observable A of subsystem 2, there is a unique observable $\tilde{A}$ of subsystem 1 defined by (4.31) which exhibits perfect correlations with A. Since, as can easily be proved,[15] the state (4.19) assumes the form

$$\psi_C = \sum_{n=1}^{\infty} |\phi_n\rangle \otimes C|\phi_n\rangle, \tag{4.33}$$

one may develop the same results with the roles of the subsystems reversed, i.e., for any observable A of subsystem 1, there exists a unique observable $\tilde{A}$ of subsystem 2 which exhibits perfect correlations with A.

An incompleteness argument similar to that given above can be given, and one can show the existence of a value map $E(O)$ on all observables of both subsystems.

4.4.2 The General Form of a Maximally Entangled State

States exhibiting ubiquitous perfect correlations are not limited to those of the form (4.19). If we examine any composite system whose subsystems are of the same dimensionality[16] and which is represented by a state[17]

Footnote 14 (Continued)

where the second equality follows from (4.22). From (4.29) follows the relationship

$$\tilde{A}_{ij} = A_{ij}^*, \tag{4.30}$$

which is a convenient form one may use to evaluate $\tilde{A}$, as we will see in Sect. 4.4.3. Note that (4.30) reduces to (4.10) when the invariant basis $\{\varphi_n\}$ of C is the position basis $|\varphi_x\rangle$..

[15] The proof of this result follows similar lines as the invariance proof given above. One expands the vectors of the state (4.19) in terms of the invariant basis $\{\varphi_n\}$. Doing so leads to an expression similar to the first equality in (4.25), with φ_i and φ_j replacing χ_i and χ_j, respectively. The delta function is then replaced using the relationship

$$\delta_{ij} = \langle\varphi_i| \left(\sum_{n=1}^{\infty} |\phi_n\rangle\langle\phi_n| \right) |\varphi_j\rangle, \tag{4.32}$$

and one can then easily develop (4.33).

[16] The derivations of this section can be carried out for an entangled system whose subsystems are either finite or infinite dimensional. Infinite sums may be substituted for the finite sums written here to develop the same results for the infinite dimensional case.

[17] This is equivalent to the form

$$\psi_{ME} = \sum_{n=1}^{N} |\psi_n\rangle \otimes |\phi_n\rangle, \tag{4.35}$$

where $\{|\psi_n\rangle\}$ is any basis of subsystem 1 and $\{|\phi_n\rangle\}$ is any basis of subsystem 2, we find that each observable of either subsystem exhibits perfect correlations with some observable of the other subsystem. To examine the properties of the state (4.35), we note that it may be rewritten[18] as

$$\psi_{ME} = \psi = \sum_{n=1}^{N} U|\phi_n\rangle \otimes |\phi_n\rangle, \tag{4.37}$$

where U is the anti-unitary operator defined by $U|\phi_n\rangle = |\psi_n\rangle$. Recall from the above discussion that (anti-linear) unitarity and anti-linearity are sufficient to guarantee both that $\{U|\phi_n\rangle\}$ is a basis if $\{|\phi_n\rangle\}$ is, and that the state shows the invariance we require. We conclude our presentation of the Schrödinger paradox by discussing this general form.

We consider a Hermitian operator A on $\mathcal{H}_N$ which can be written as

$$A = \sum_{n=1}^{N} \mu_n |\phi_n\rangle \langle \phi_n|. \tag{4.38}$$

We then define the observable $\tilde{A}$ by[19]

$$\tilde{A} = U A U^{-1}. \tag{4.40}$$

Footnote 17 (Continued)

$$\psi_{ME} = \sum_{n=1}^{N} c_n |\psi_n\rangle \otimes |\phi_n\rangle, \tag{4.34}$$

where $\{|\psi_n\rangle\}$ is any basis of subsystem 1 and $\{|\phi_n\rangle\}$ is any basis of subsystem 2 and the $|c_n|^2 = 1 \ \forall n$.

[18] At this point, one may address the objection that the EPR state (4.8) makes the positions of the two particles coincide. For example, one can consider an anti-unitary operator U_d defined by

$$U_d|\psi\rangle = \int_{\infty}^{\infty} dx |\varphi_x\rangle\langle\psi|\varphi_{x+d}\rangle, \tag{4.36}$$

where d is an arbitrary constant. Then, in the state (4.37), the two particles are separated by a distance d.

[19] For comparison with the form (4.10), and (4.30), we note that if we express the operator U as $U = C\bar{U}$ with C an anti-unitary involution, and $\bar{U}$ a unitary matrix, then (4.40) leads to

$$\tilde{A}_{ij} = (\bar{U} A \bar{U}^{-1})^*_{ij}, \tag{4.39}$$

where the ij subscript indicates the ijth matrix element in terms of $\{\varphi_n\}$, the basis that is invariant under C.

For any given A, this defines a unique operator $\tilde{A}$. Using $A|\phi_n\rangle = \mu_n|\phi_n\rangle$ and $UU^{-1} = 1$, we have

$$UAU^{-1}U|\phi_n\rangle = \mu_n U\phi_n, \tag{4.41}$$

so that the eigenvectors and eigenvalues of $\tilde{A}$ are given by $\{U\phi_n\}$ and $\{\mu_n\}$.

Identifying A as an observable of subsystem 2 and $\tilde{A}$ as an observable of subsystem 1, we can see from the form of (4.37) that these exhibit perfect correlations for such a state. As in the EPR case and its generalization, the basis-invariance and symmetry of the state imply that for *any* observable A of either subsystem, there is a unique observable $\tilde{A}$ of the other with which A is perfectly correlated. An argument similar to that given for the EPR state leads to the existence of a non-contextual value map $E(O)$ on all observables of subsystem 1 and all observables of subsystem 2.

4.4.3 Maximally Entangled State with Two Spin-1 Particles

To make the principles under consideration more plain, it is instructive to consider in detail a particular maximally entangled state. For this purpose, we select an entangled state of two spin-1 particles. The selection[20] is not arbitrary, but will serve as a useful reference when we come to discuss the Conway and Kochen Free Will theorem in Sect. 4.5.2.

Let the basis vectors of a spin-1 particle be represented by $|+1\rangle$, $|0\rangle$ and $|-1\rangle$ corresponding to eigenvalues 1, 0 and -1, respectively, for the z-component of the spin. Suppose the system is a two particle state represented by

$$\psi = |+1\rangle \otimes |-1\rangle + |-1\rangle \otimes |+1\rangle - |0\rangle \otimes |0\rangle, \tag{4.42}$$

where the normalization has been suppressed for simplicity.

First, we would like to write out the explicit matrix formulation of U appearing in Eq. 4.37 such that this reduces to (4.42). Second, because the Kochen and Specker observables are of central importance to the Free Will Theorem, we also wish to analyze how perfect correlations operate in the case of the set $\{S^2_{\theta,\phi}\}$. I.e., for two spin-1 particles represented by (4.42), we wish to determine which observables $\{\widetilde{S^2_{\theta,\phi}}\}$ of one particle will exhibit perfect correlations with the set $\{S^2_{\theta,\phi}\}$ of the other.

As we saw in the previous section, the maximally entangled states are characterized by an anti-unitary operator U. In terms of U, the state can be written as in Eq. 4.37 which we repeat here for convenience:

$$\psi_{ME} = \psi = \sum_{n=1}^{N} U|\phi_n\rangle \otimes |\phi_n\rangle. \tag{4.43}$$

---

[20] The state considered here is a special case of the general class of three-dimensional maximally entangled state (4.52) to be discussed in Sect. 4.5.1. In the discussion given there we will develop a proof of quantum nonlocality for this class of states and some others as well.

The perfect correlations between particles are then developed using the Eq. 4.40, according to which the observable $\tilde{A}$ of one particle or subsystem will show perfect correlations with A of the other particle or subsystem such that $\tilde{A} = UAU^{-1}$.

Inspecting the form given in Eq. 4.34, it is quite evident that (4.42) is a maximally entangled state. What is needed is the appropriate anti-unitary operator U for agreement with (4.43). We write $U = C\bar{U}$, where $\bar{U}$ is a unitary operator given by

$$\bar{U} = \begin{pmatrix} 0 & 0 & 1 \\ 0 & -1 & 0 \\ 1 & 0 & 0 \end{pmatrix}$$

and C is an anti-unitary involution under which the S_z eigenvectors are invariant, i.e., $C|+1\rangle = |+1\rangle$, $C|-1\rangle = |-1\rangle$ and $C|0\rangle = |0\rangle$ Using these eigenvectors as the basis in the maximally entangled state (4.37), we obtain

$$\psi = U|+1\rangle \otimes |+1\rangle + U|0\rangle \otimes |0\rangle + U|-1\rangle \otimes |-1\rangle \tag{4.44}$$

A trivial matrix multiplication yields

$$\psi = |+1\rangle \otimes |-1\rangle + |-1\rangle \otimes |+1\rangle - |0\rangle \otimes |0\rangle, \tag{4.45}$$

which is just the form (4.42) given above. Therefore, the anti-unitary operator U where $U = C\bar{U}$ defined above is the correct operator to define our state (4.42) as a maximally entangled state.

Next, we evaluate $US_{\theta,\phi}^2 U^{-1}$. To begin, we recall from elementary quantum mechanics the form of the operator $S_{\theta,\phi}$:

$$S_{\theta,\phi} = \begin{pmatrix} \cos\theta & \frac{1}{\sqrt{2}}\sin\theta e^{-i\phi} & 0 \\ \frac{1}{\sqrt{2}}\sin\theta e^{i\phi} & 0 & \frac{1}{\sqrt{2}}\sin\theta e^{-i\phi} \\ 0 & \frac{1}{\sqrt{2}}\sin\theta e^{i\phi} & -\cos\theta \end{pmatrix}. \tag{4.46}$$

To determine $S_{\theta,\phi}^2$ requires simple matrix multiplication yielding:

$$S_{\theta,\phi}^2 = \frac{1}{2}\begin{pmatrix} \cos^2\theta + 1 & \frac{1}{\sqrt{2}}\sin 2\theta e^{-i\phi} & \sin^2\theta e^{-2i\phi} \\ \frac{1}{\sqrt{2}}\sin 2\theta e^{i\phi} & 2\sin^2\theta & -\frac{1}{\sqrt{2}}\sin 2\theta e^{-i\phi} \\ \sin^2\theta e^{2i\phi} & -\frac{1}{\sqrt{2}}\sin 2\theta e^{i\phi} & \cos^2\theta + 1 \end{pmatrix}. \tag{4.47}$$

Using U as defined above, we must evaluate $US_{\theta,\phi}^2 U^{-1}$. Using $C = C^{-1}$, we find that

$$US_{\theta,\phi}^2 U^{-1} = C\bar{U}S_{\theta,\phi}^2\bar{U}^{-1}C. \tag{4.48}$$

Simple matrix multiplication yields

$$\bar{U}S^2_{\theta,\phi}\bar{U}^{-1} = \frac{1}{2}\begin{pmatrix} \cos^2\theta + 1 & \frac{1}{\sqrt{2}}\sin 2\theta e^{i\phi} & \sin^2\theta e^{2i\phi} \\ \frac{1}{\sqrt{2}}\sin 2\theta e^{-i\phi} & 2\sin^2\theta & -\frac{1}{\sqrt{2}}\sin 2\theta e^{i\phi} \\ \sin^2\theta e^{-2i\phi} & -\frac{1}{\sqrt{2}}\sin 2\theta e^{-i\phi} & \cos^2\theta + 1 \end{pmatrix}. \qquad (4.49)$$

It remains only to calculate the expression CMC, substituting the right hand side of (4.49) for M. In Sect. 4.4.1, we saw that an expression of the form CMC can be evaluated in terms of its matrix elements in $\{\varphi_n\}$, the basis that is invariant under C. To do so, we use the relation (4.29):

$$(CMC)_{ij} = M^*_{ij}. \qquad (4.50)$$

Using this relationship together with (4.49) and (4.48), it is easy to show that

$$US^2_{\theta,\phi}U^{-1} = \frac{1}{2}\begin{pmatrix} \cos^2\theta + 1 & \frac{1}{\sqrt{2}}\sin 2\theta e^{-i\phi} & \sin^2\theta e^{-2i\phi} \\ \frac{1}{\sqrt{2}}\sin 2\theta e^{i\phi} & 2\sin^2\theta & -\frac{1}{\sqrt{2}}\sin 2\theta e^{-i\phi} \\ \sin^2\theta e^{2i\phi} & -\frac{1}{\sqrt{2}}\sin 2\theta e^{i\phi} & \cos^2\theta + 1 \end{pmatrix}. \qquad (4.51)$$

Comparing this result with $S^2_{\theta,\phi}$, we have that $US^2_{\theta,\phi}U^{-1} = S^2_{\theta,\phi}$.

At this point, we have sufficient information to address our original inquiry. Evidently, for the Kochen and Specker observables $\{S^2_{\theta,\phi}\}$ defined relative to one particle, the perfectly correlated observables $\{\widetilde{S^2_{\theta,\phi}}\}$ of the second particle are simply the *same observables*. Thus, for two particles defined by the state (4.42), a measurement of $S^2_{\theta,\phi}$ on one particle always gives a result equal to that obtained in measurement of same observable $S^2_{\theta,\phi}$ of the other particle.[21]

4.5 Schrödinger Nonlocality

In the previous chapter we examined the Einstein–Podolsky–Rosen paradox, reviewing the essential mathematics and the physical implications.[22] We were able to identify basic assumptions present in the analysis, and define a clear conclusion.

---

[21] A very similar proof can be given for the spin-singlet state discussed in Chap. 3 in the presentation of the Einstein–Podolsky–Rosen paradox. By such a proof, one may show that the spin singlet state is a two-dimensional maximally entangled state with the anti-unitary operator $U = C\bar{U}$, where $\bar{U}$ is unitary and is given by $\bar{U} = \begin{pmatrix} 0 & 1 \\ -1 & 0 \end{pmatrix}$. The operator C is an anti-unitary involution under which the σ_z eigenvectors are invariant, i.e., $C|\uparrow\rangle = |\uparrow\rangle$ and $C|\downarrow\rangle = |\downarrow\rangle$. Using this form for U, one can easily show that the perfect correlations turn out just right, as well.

[22] Our study was addressed to the Bohm version of the Einstein–Podolsky–Rosen paradox, so that the relevant observables are spin components, rather than position and momentum. As we have shown in the present chapter, an analogous conclusion follows for the original EPR formulation as well. Furthermore, we repeated Schrödinger's analysis and developed the fact the EPR quantum state exhibits similar correlations for all observables of the two particles, and not only position and momentum.

With this, it became possible to show that the Einstein–Podolsky–Rosen paradox is equivalent to a theorem. By the Einstein–Podolsky–Rosen Theorem, the assumptions of locality with perfect correlations on spin-$\frac{1}{2}$ observables, $\sigma_{\theta,\phi}^{(1)}$ and $\sigma_{\theta,\phi}^{(2)}$, led to the conclusion of noncontextual hidden variables on these quantities.

Further, we discussed the relationship between the Einstein–Podolsky–Rosen analysis and Bell's Theorem. Having understood the EPR Theorem, it became clear that the argument could be combined with Bell's Theorem to create what is known as a "complex argument." (see Sect. 3.4). From this, we demonstrated what follows is a disagreement between the predictions of quantum mechanics and those of any local theoretical construction.

In this chapter we examined Erwin Schrödinger's contribution to the issue. We saw that Schrödinger extended the work of EPR, as he showed that the quantum state of EPR implies perfect correlations in a far broader class of observables and indeed, for all observables associated with each particle. As was the case for the EPR paradox, we were able to show that Schrödinger's result is also equivalent to a theorem.

Being aware that the Einstein–Podolsky–Rosen paradox leads to such a powerful result when combined with Bell's Theorem, one might naturally be curious about whether something similar follows from the Schrödinger paradox. What we are seeking is any mathematical theorem which bears the same relationship to Schrödinger paradox that Bell's Theorem takes to the Einstein–Podolsky–Rosen paradox.

As it happens, what is required is precisely the type of theorem already discussed in Chap. 2. The relevant category of argument are the "no-hidden-variables" proofs which, like Gleason's and Kochen and Specker's theorem, address the question of a noncontextual value map on the observables. In Chap. 2, we showed in Sect. 2.4 that Gleason's theorem as well as Kochen and Specker's may be understood as falling into a category known as "spectral incompatibility theorems."

4.5.1 Schrödinger Paradox and Spectral Incompatibility Theorems

Chapter 2 of the present book was devoted to the issue of contextuality and the no-hidden-variables theorems due to Gleason, and Kochen and Specker. The latter are mathematical arguments showing that certain forms of mappings from observables to values are "impossible" in that they cannot replicate the predictions of quantum mechanics. As mathematical arguments, there is nothing objectionable in these results. It is in the claims made regarding their physical implications, i.e., the impossibility of determinism in quantum mechanics, that the authors fall into error. Remembering the principle of contextuality, the limitations of the theorems in question becomes clear. Physically, Gleason's theorem and Kochen and Specker's theorem serve only to disprove noncontextual hidden variables.[23]

---

[23] An accomplishment which does not in itself constitute a significant restriction on quantum realism, since contextuality is a feature that flows naturally from the quantum formalism itself.

What is the relationship between these theorems and the Schrödinger paradox? Recall that in the above discussion of Sect. 4.3.2, we demonstrated that Schrödinger's argument is equivalent to a theorem. Schrödinger's Theorem tells us the assumptions of locality together with the ubiquitous and context-free perfect correlations we must conclude noncontextual hidden variables on the observables. The link to Gleason's theorem and to Kochen and Specker's theorem is immediate. Each of these begins with the conjecture that such theoretical a structure—noncontextual variables—is possible.

While the connection between Schrödinger paradox and these theorems is simple and direct, the consequences are dramatic. When the theorems are taken in conjunction,[24] one finds that the only conclusion possible is that the predictions of quantum mechanics cannot be replicated by any local theory.[25]

Let us first consider the state of affairs that arise from the Kochen and Specker theorem and its relation to Schrödinger's Theorem. Suppose a system is described by a maximally entangled state[26] whose subsystems are of dimensionality three:

$$\psi = \sum_{n=1}^{3} \phi_n \otimes \psi_n. \tag{4.52}$$

For this state,[27] the individual particles' observables are each on three-dimensional spaces, which we designate as $\mathcal{H}_1$ and $\mathcal{H}_2$. Now, this state is clearly one of the maximally entangled states, as we discussed in Sect. 4.4. Therefore, there are perfect correlations between every observable A^3 associated with particle-1 and some observable $\tilde{A}^3$ of particle-2,[28] where the superscript 3 reflects that the observable is defined on a three-dimensional space. By the Schrödinger paradox results derived in this chapter, it follows that for two-particle state described by Eq. 4.52, there exist noncontextual hidden variables on the sets $\{A^3\}$ and $\{\tilde{A}^3\}$ associated with particle-1 and particle-2.

---

[24] I.e., when one combines Schrödinger paradox with either Gleason's or Kochen and Specker's theorem, one finds such a result.

[25] Similar nonlocality results were discovered by Brown and Svetlichny [1], Heywood and Redhead [2], Aravind [3], and by Cabello [4].

[26] This is the generalization the state (4.42) examined in Sect. 4.4.3. The aims of the discussion were different in that prior section, where we wished to develop an explicit example of a maximally entangled state. As was mentioned there, that particular state is of great relevance to the Conway and Kochen Free Will theorem. In the present section, we are interested in deriving the Schrödinger nonlocality based upon the Schrödinger paradox in conjunction with Kochen and Specker's theorem. Such instances of quantum nonlocality are not restricted to the state (4.42), but can be derived from the more general form here, and also for infinite dimensional cases, as we shall see.

[27] Of course, such a system need not consist of two spin-1 particles. If it does not, then the link between Schrödinger paradox and the Kochen and Specker theorem may be made by considering those observables which are formally equivalent to the sets $\{S^2_{\theta,\phi}\}$ and $\{\tilde{S}^2_{\theta,\phi}\}$. Since any two Hilbert spaces of the same dimension are isomorphic, it follows that such formally equivalent observables must exist.

[28] Which we developed explicitly for a special case, in Sect. 4.4.3.

Consider now Schrödinger's Theorem.[29] For the current situation, it will reduce to the form: if one assumes locality together with the noncontextual perfect correlations on all A^3 of particle-1 and all $\tilde{A}^3$ of particle-2, then there must exist (noncontextual) hidden variables on these observables.

In Chap. 2, we discussed the Kochen and Specker theorem. This argument concerns a mapping $E(O)$ from quantum observables O to values E, in particular, a function on the squares of the various components of a spin-1 particle, $\{S^2_{\theta,\phi}\}$. In quantum mechanics, the values of such observables are given by the eigenvalues $0,1$, and we have also that for any orthogonal directions in space x,y,z the values must satisfy $E(S^2_x) + E(S^2_y) + E(S^2_z) = 2$. By an intricate geometrical argument,[30] Kochen and Specker prove that it is a mathematical impossibility for any function $E(O)$ on the spin observables in question to meet these quantum criteria.

We can trvially reexpress Kochen and Specker's result into the following form. Given any one-to-one value-map $E(O)$ on the spin-1 components, it is impossible to meet the quantum mechanical requirement that every commuting set is assigned a joint-eigenvalue.

As we recall from the previous chapter, when one has two arguments such that conclusion of one is identical to a premise of the second, their conjunction forms a "complex argument." Relative to the second argument, the conclusion of the first is known as a *non-basic premise*. If the resulting complex argument is valid (which follows from the validity of each component), then the basic premises of the two arguments imply the conclusion of the second.

From what we have seen, it is clear that Schrödinger's Theorem may be combined[31] with Kochen and Specker's theorem to form a complex argument. Within this complex argument, we find that *the existence of noncontextual hidden variables is a nonbasic premise of the Kochen and Specker theorem component.*

We may assert that the resulting complex argument is valid due to the fact that each of its components are theorems. If we look to each of the theorems to gather the basic premises, we find from Schrödinger's Theorem locality together with the universal, context-free perfect correlations. In Kochen and Specker's theorem[32] the only assumption is the noncontextul value-map $E(\{S^2_{\theta,\phi}\})$. However, within the complex argument this assumption is non-basic. The conclusion of the overall argument is simply that of the Kochen and Specker theorem, namely that the empircal predictions cannot be made to agree with the joint-eigenvalues required by quantum theory. In total, we have this structure: given locality, together with the Schrödinger perfect

---

[29] Discussed in Sect. 4.3.2. Note that the existence of this theorem implies a rather broad result, namely that for *any* theory which meets the assumptions, the conclusion must follow.

[30] We presented some of the details in Chap. 2.

[31] Some readers may object that the conclusion of Schrödinger's Theorem is not identical to the value map assumption made by Kochen and Specker. However, it is trivial to show that the former implies the latter: if one has a value map on all observables (of either particle) then certainly this entails the existence of a map on any *subset* of those observables, including the set $\{S^2_{\theta,\phi}\}$.

[32] For our purposes, the logic becomes more streamlined if we rearrange Kochen and Specker's theorem as "Assuming the noncontextual value map $E(\{S^2_{\theta,\phi}\})$ a mapping to joint-eigenvalues is impossible".

correlations we are led to a contradiction with the quantum joint-eigenvalues for spin-1 particles.

Let us examine this conclusion carefully. Note that the assumptions of the argument—locality and perfect correlations—leave room for a variety of hypothetical physical theories. Therefore, when we consider the description of the maximally entangled state (4.52), what we must conclude is that any local theory exhibiting the Schrödinger perfect correlations must contradict the quantum mechanical predictions. It follows logically, using the same argument as presented in Chap. 3, that the quantum mechanical predictions for this state cannot be replicated by any local theory.

These considerations were made in the context of a three-dimensional maximally entangled state, given in (4.52). However, if we consider any maximally entangled state of dimensionality N of at least three:

$$\psi = \sum_{n=1}^{N} \phi_n \otimes \psi_n, \tag{4.53}$$

then we can evidently develop a nonlocality proof for this state by using the Kochen and Specker theorem. To see this, we re-write (4.53) as

$$\psi = \phi_1 \otimes \psi_1 + \phi_2 \otimes \psi_2 + \phi_3 \otimes \psi_3 + \sum_{n=4}^{N} \psi_n \otimes \psi_n, \tag{4.54}$$

and we define $\mathcal{H}_2$ to be the subspace of particle-2's Hilbert space spanned by the vectors ψ_1, ψ_2, ψ_3, and the operator P as the projection operator of $\mathcal{H}_2$. Let us define the set of observables $\{\zeta_{\theta,\phi}\}$ such that they are formally identical to the squares of the spin components $\{\zeta_{\theta,\phi}\}$ on $\mathcal{H}_2$ and give zero when operating on any vector in its orthogonal complement. Then the Kochen and Specker theorem tells us that the existence of definite values for the set $\{P, \{\zeta_{\theta,\phi}^2\}\}$ must conflict with quantum mechanics. To see this, consider those joint-eigenvalues of $\{P, \{\zeta_{\theta,\phi}\}\}$ for which $P = 1$. The corresponding values of the set $\{\zeta_{\theta,\phi}\}$ in this case are equal to joint-eigenvalues on the spin observables themselves. Since the Kochen and Specker theorem implies the impossibility of an assignment of values to the spin components, then the same follows for the observables $\{\zeta_{\theta,\phi}\}$ given that $P = 1$.

Considering these results with the Schrödinger paradox implications for the state (4.53),[33] one again finds a proof of the empirical disagreement between quantum theory and all local theoretical explanations.

If we analyze the conjunction of Schrödinger's paradox with Gleason's theorem, we find another such result follows. From Gleason's theorem we have that a value map on the set of all projections $\{P\}$ must conflict with quantum mechanics. One can show that for any maximally entangled state

---

[33] The necessary existence of the noncontextual values on all observables including $\{P, \{\zeta_{\theta,\phi}\}\}$.

$$\sum_{n=1}^{N} \phi_n^{(1)} \otimes \psi_n^{(2)}, \tag{4.55}$$

where N is at least three, we have quantum nonlocality. This result also holds true for maximally entangled states of infinite-dimensionality, in which case the sum in (4.55) is replaced by an infinite sum.

Thus, for all maximally entangled states of dimensionality at least 3, one may develop such a quantum nonlocality proof from either Gleason's or Kochen and Specker's theorem in conjunction with Schrödinger's Theorem. We have seen that these proofs differ from that involving Bell's theorem in several ways. First, the Schrödinger nonlocality is of a deterministic, rather than statistical character. This is due to the fact that the conflict of incompleteness with quantum mechanics is in terms of the quantum prediction for individual measurements, i.e., the prediction that measurements of a commuting set always give one of that set's joint-eigenvalues. Second, such quantum nonlocality proofs can be developed from these theorem's implications regarding the observables of just *one* subsystem. Finally, in some instances the Schrödinger nonlocality may be proved for a larger class of observables than can the EPR/Bell nonlocality.

It is important to note that the above discussions represent just two of a large family of potential new instances of quantum nonlocality. The Schrödinger paradox results in a powerful conclusion, namely a value map on all observables of each particle described by the maximally entangled state. This effectively opens a broad range of possibilities to construct the sort of complex argument described above. All one requires is an argument which contradicts such a wholesale mapping of predetermined values to observables. It should be clear that any spectral incompatibility theorem (see Sect. 2.4) will suffice for this purpose. If the theorem in question concerns a class of observables on an N-dimensional Hilbert space, then the proof can be applied to any maximally entangled states whose subsystems are at least N-dimensional. We suggest the appellation "Schrödinger nonlocality" for such results.

4.5.2 What is Proven by the Conway–Kochen Free Will Theorem

The Conway Kochen Free Will Theorem[34] was advanced as an argument demonstrating the impossibility of deterministic interpretations as a description of nature. In particular, the authors maintain they have created a secure link between the impossibility of deterministic hidden variables and the existence of human free will. "... we prove that if the choice of a particular type of spin 1 experiment is not a function of the information accessible to experimenters, then its outcome is equally not a function of the information accessible to the particles." Emphasizing that their proof rests

---

[34] Please see Conway and Kochen in [13] and also in [14].

on very modest axioms, they claim to have arrived at a correspondingly irrefutable argument. With the insights we have developed in our efforts so far, both the Einstein–Podolsky–Rosen argument and Schrödinger's paradox, it is not difficult to see why the Free Will Theorem falls short of the authors' assertions.[35] The fallacy[36] in their argument is not that the assumptions are unreasonable, but rather the opposite, i.e., that said axioms are more potent than their authors acknowledge, such that a powerful conclusion follows from these without further conceptual additions. In light of the Schrödinger paradox and its implications, it becomes clear that Conway and Kochen's assumptions of deterministic hidden variables and the human free will are quite superfluous to their argument. Furthermore, the conclusion of their argument's main component[37] must be regarded as a half-truth, in that a stronger conclusion actually follows from the three axioms they stress. As we proceed to correct these shortcomings, it will become clear that the structure with which Conway and Kochen are operating is in fact a special case of Schrödinger nonlocality, namely the very same which we presented above when discussing the state (4.52). As such, the Free Will Theorem offers a mathematical construction which, when correctly interpreted, does not restrict the possibility of determinism, but constitutes another manifestation of the conflict between quantum mechanics' empirical predictions with any local theoretical explanation.

Asking readers to consider a system of two spin-1 particles, Conway and Kochen present their argument, offering simple assumptions and moving through their steps in a logical and straightforward manner. In this section, we shall first briefly review Conway and Kochen's argument and then point out where the authors have overlooked certain crucial facts, and how the oversight alters the implications of the analysis.

First, the Conway and Kochen analysis[38] sets up the axioms "SPIN" "TWIN," and "FIN." SPIN is essentially the familiar characteristics of measurement of a spin-1 particle. In particular, according to SPIN, there is an operation known as measurement of the square of a component of spin, and when a measurement is performed on any orthogonal triplet S_x^2, S_y^2, S_z^2 the results are the numbers from the set 1, 0, 1, in some order. Very clearly, this axiom differs in no essential way with the quantum mechanical prescription for a spin-1 particle, as mentioned in Chap. 2,

---

[35] As we will see, the axioms of Conway and Kochen are identical to those of the particular case of Schrödinger's analysis for a maximally entangled state of two spin-1 particles (Eq. 4.42) discussed in Sect. 4.4.3. Therefore, the conclusions developed in Sect. 4.5.1 for the more general quantum state of Eq. 4.52 must apply.

[36] Others who raise similar objections to Conway and Kochen are Bassi and Ghirardi [5], Tumulka [6] and also Goldstein et al. [7].

[37] The Conway Kochen Free Will theorem consists of two components. What one might refer to as the main component, is that in which the axioms SPIN, TWIN and FIN are called on to develop the conclusion of noncontextual hidden variables. As we will see, this is essentially nothing different than a special case of the Schrödinger paradox, except that certain facts of the analysis are overlooked. The second component is the Kochen and Specker theorem.

[38] We proceed by first considering the original form of the "Free Will Theorem". The "Strong Free Will Theorem" will be dealt with afterwards.

Sect. 2.2. From quantum mechanics we have that the eigenvalues of the squares of the spin-components of a spin 1 particle are 0, 1, and the measurement results are constrained such that $S_x^2 + S_y^2 + S_z^2 = 2$, for any orthogonal triplet x, y, z. Nor do the authors themselves consider this as a departure from quantum theory: "...no physicist would question the truth of our SPIN axiom, since it follows from quantum mechanics."[39]

Secondly we have the "TWIN" axiom asserting that measurements on the two particles by independent laboratory experimenters A and B always exhibit correlations of the following type. If A measures the components aligned with the triplet x, y, z, i.e., observables S_x^2, S_y^2, S_z^2 on one particle and B measures the component in the direction w (S_w^2) where w is one of the x, y, z, then the value of S_w^2 will be the same as the corresponding result for the matching observable of S_x^2, S_y^2, S_z^2 obtained by experimenter A.

According to the FIN axiom, no information may be transmitted faster than light speed.

The Conway and Kochen paper continues from this point by stating that the SPIN, TWIN and FIN assumptions are sufficient to prove their main theorem: "If the choice of directions in which to perform spin-1 experiments is not a function of the information available to experimenters, then the responses of the particles are equally not functions of the information available to them."

In the first part of their discussion, the authors review the Kochen and Specker theorem[40] and its implications towards deterministic interpretations of quantum mechanics. The notation is $\theta(w)$ is introduced to represent the result of a spin-squared component measurement performed in direction w. With this, the possibility of interpreting the experimental results in a noncontextual way is immediately eliminated, and the value map is shown "impossible" in the sense that it cannot satisfy the (quantum mechanical) relationship mentioned in SPIN above, i.e., that the values are 0,1 and for each orthogonal set they must sum to 2.

Next the authors attempt to more generally address deterministic formulations of quantum mechanics. In developing the problem they write $\theta_a(x, y, z; \alpha')$ for the result of an experiment measuring particle a's spin-squared components in orthogonal directions x, y, z. According to the authors, the variable α' is meant to allow for some additional information available to the particle. The expression $\theta(x = ?, y, z; \alpha') = i$ where $i = 0, 1$ is the form Conway and Kochen offer to represent the measurement result. They give similar expressions for the other members of the triplet.[41]

For the second particle, labelled b, they consider functions $\theta_b(w; \beta')$ which map every direction w to a measurement result for the experiment performed by B to

---

[39] See [14].

[40] More particularly, the version of A. Peres. See Peres [15]. For further information on the Kochen and Specker theorem, see Chap. 2.

[41] Note that this formulation allows for the value of each observable to depend upon the measurement context. E.g., an experiment to determine S_x^2 when measured as part of the orthogonal triplet S_x^2, S_y^2, S_z^2 would be represented by $\theta(x = ?, y, z; \alpha')$. Similar comments apply for S_y^2 and S_z^2.

determine that spin-squared component. As one might guess, the second variable β' refers to the information available to the particle b.

The twin axiom implies that there is equality between the results obtained by experimenter B measuring S_w^2 and A measuring the set S_x^2, S_y^2, S_z^2 if the axis w is equal to one of x, y, z. Using the notion presented above, this is expressed mathematically as

$$\theta_b(w, \beta') = \begin{cases} \theta_a(x =?, y, z; \alpha') & \text{if } w = x \\ \theta_a(x, y =?, z; \alpha') & \text{if } w = y \\ \theta_a(x, y, z =?; \alpha') & \text{if } w = z \end{cases} \tag{4.56}$$

If we insist on the free will of each of the two experimenters, then for at least some values α' and β', the functions in Eq. 4.56 must exist for all directions w, x, y, z. This follows from the fact that there is no restriction upon the experimenters' choice of measurement axes.

It also follows that for whatever α' and β' the functions are defined, the measurement axes x, y, z cannot depend upon the information variable α', and the axis w cannot depend upon β'. One can appeal to the free choice of the experimenters to guarantee this independence. Moreover, it is clear from FIN that x, y, z does not depend upon β' and w does not depend on α'.

The variables α' and β' are designed to represent whatever information, apart from x, y, z and w, is available to each of a and b, respectively. We have already established that there must exist some particular values α_0 and β_0 for which these functions are defined for all x, y, z and w.

Note that if we choose a fixed values α_0, β_0 of the information variables, then the following consequences follow. First, for the function $\theta_b(w; \beta_0)$ becomes mapping from directional choice w to measurement results.[42] As such, we may define a new function $\theta_0(w) \equiv \theta_b(w; \beta_0)$.

Returning to Eq. 4.56, we find that this becomes

$$\theta_0(w) = \begin{cases} \theta_a(x =?, y, z\alpha_0) & \text{if } w = x \\ \theta_a(x, y =?, z; \alpha_0) & \text{if } w = y \\ \theta_a(x, y, z =?; \alpha_0) & \text{if } w = z \end{cases} \tag{4.57}$$

However, if the function on the left hand side depends only on direction, then the same must be true for functions on the right, so that we must have:

$$\theta_0(w) = \begin{cases} \theta_0(x) & \text{if } w = x \\ \theta_0(y) & \text{if } w = y \\ \theta_0(z) & \text{if } w = z \end{cases} \tag{4.58}$$

Therefore, for the particle a as well, the argument has led us to the notion that deterministic formulations are restricted to the form of a one-to-one mapping from spin-squared components to their values.

---

[42] Such a mapping clearly makes no allowance for measurement procedure $\mathcal{M}(S_w^2)$ which B brings to bear, so that the Kochen and Specker theorem comes into play.

At this point, it is only necessary to recall that Kochen and Specker theorem, which denies the possibility of such a mapping. By the Kochen and Specker theorem, any attempt at such a mapping which hopes to concur with the quantum mechanical predictions cannot succeed.

In summary, the Free Will Theorem takes the following form. The quantum predictions for a spin-1 particle are incorporated into the argument in the form of the axiom SPIN. TWIN then presumes a two-particle system on which the values of the squares of various components of the spin of always exhibit equal results in distant laboratories, independent of the selection of measurement procedure. Finally, FIN prescribes that no information may be transmitted between separate locations faster than light speed. These axioms are taken to be minimal and uncontroversial, and are thereby considered to lead to a very robust conclusion. When one considers the possibility of deterministic representations of quantum mechanics, hence functions of the form $\theta(x, y, z; \alpha)$ which determine measurement results, in conjunction with human free will, the result is that the only possible functions are simple one-to-one mappings from the spin-squared components to their values, e.g., $\theta(w)$. The latter possibility is known to conflict with quantum mechanics via the Kochen and Specker theorem.

Thus, it would appear that according to the Free Will Theorem, if one admits SPIN, TWIN and FIN, along with human free will then any suggestion of deterministic interpretations of quantum mechanics must fail. However, a brief examination of these axioms combined with our understanding of the Schrödinger paradox uncovers the flaw of Conway and Kochen's argument and leads us to reassess this conclusion.

4.5.2.1 Free Will Theorem and Schrödinger Nonlocality

To see where Conway and Kochen's argument goes astray, let us begin by looking at the broad structure of their efforts. It is clear from the above that the argument consists of two main components, one being the classic Kochen and Specker theorem, the other being the argument in which the by now familiar SPIN, TWIN and FIN play a direct role. The latter argument begins with these axioms and attempts to address the question of a deterministic formulation of quantum mechanics, as given by mathematical functions $\theta_a(x, y, z; \alpha')$. Besides these concepts, the only further assumption they require is that of human free will. They are then able to arrive at the conclusion given by Eq. 4.58, according to which the deterministic formulation must reduce to a one-to-one (i.e., noncontextual) mapping $\theta_0(w)$ of the spin observables to their values.

It is not difficult to see that this analysis (perhaps one may call it the SPIN–TWIN–FIN argument) is equivalent[43] to the Schrödinger paradox example based on

---

[43] Some might claim that Schrödinger paradox and Schrödinger's Theorem themselves depend upon the assumption of human free will. We discussed this issue for the EPR/Bell case in Sect. 3.4.3. There we showed that in addition to locality and perfect correlations, the analysis assumes only freedom from superdeterminism. This latter is much weaker than human free will and it is a

the quantum state (4.52), presented above.[44] With such an identification, it is clear that when Conway and Kochen combine the SPIN–TWIN–FIN argument with the Kochen and Specker theorem, they are proceeding in essentially the same manner as we have done above, i.e., to form a complex argument whose overall implication is quantum nonlocality. We now proceed to make this explicit and thereby demonstrate that such is the true implication of the Free Will Theorem.

We compare the SPIN–TWIN–FIN argument to the Schrödinger paradox example related to the state (4.52). To do so, let us reconsider the three basic axioms. The TWIN axiom implies the correlation between measurement results obtained by different experimenters A and B examining spin-squared components of a spin 1 particles. If A measures the components aligned with the triplet x, y, z, i.e., observables S_x^2, S_y^2, S_z^2 and B measures the component in the direction w (S_w^2) where w is one of the x, y, z, then the value of S_w^2 will be the same as the corresponding result for the matching observable of S_x^2, S_y^2, S_z^2 obtained by experimenter A.

Examination of the TWIN axiom shows that such correlations are identical to the Schrödinger paradox perfect correlations for the maximally entangled state (4.42) discussed in Sect. 4.4.3. In Sect. 4.4, we discussed the form of maximally entangled states and developed a relationship between those observables of the two particles or systems which exhibit perfect correlations. We showed that every such state may be written as (4.37) and that the perfectly correlated observables are related to one another via the relation $\tilde{A} = UAU^{-1}$. In the special case of two spin-1 particles expressed by (4.42), we demonstrated the set of observable $S_{\theta,\phi}^2$ of particle-1 will be perfectly correlated with the *same observables* of the other, i.e., we showed that $\widetilde{S_{\theta,\phi}^2} = S_{\theta,\phi}^2$.

Let us remember also that in discussing the Schrödinger paradox, we noted that the 100% agreement between observables holds true *independently* of the choice of measurement procedure (see Sect. 4.3). In effect, no matter which measurement $\mathcal{M}(A)$ is used to find the value of A, there is perfect correlation with the result obtained from measurement of the corresponding observable $\tilde{A}$ of the partner particle. Thus the Schrödinger paradox perfect correlations for the state (4.42) match all the characteristics of the TWIN axiom.

According to the FIN axiom, no information may be transmitted faster than light speed. Such a requirement also appears among the assumptions of the Schrödinger paradox.[45]

The SPIN axiom, as we have mentioned, contains nothing beyond what quantum mechanics itself prescribes for spin 1 particles. In particular, according to SPIN,

---

Footnote 43 (Continued)

fundamental assumption behind all science. By essentially the same arguments given in Sect. 3.4.3, one can demonstrate that it is equally incorrect to assert that the Schrödinger analysis depends upon human free will.

[44]   In fact, the Conway and Kochen argument may be derived from the maximally entangled state discussed in Sect. 4.4.3, namely that represented by (4.42), which is a special case of the form (4.52)

[45]   We demonstrated that the Schrödinger paradox depends upon locality in the discussion of Schrödinger's Theorem presented in Sect. 4.3.2.

there is an operation known as measurement of the square of a component of spin, and when a measurement is performed on any orthogonal triplet S_x^2, S_y^2, S_z^2 the results are the numbers from the set 1,0,1, in some order.

How does this axiom relate to what is taking place in the Schrödinger paradox? We demonstrated in Sect. 4.3.2 that Schrödinger's analysis is equivalent to a theorem. In this way, the logical structure implied by the paradox was brought forth. However, we did not explicitly state which aspects of quantum mechanics are required as background assumptions in this analysis. In order to connect the Schrödinger paradox with the Free Will Theorem, this question must now be addressed.

At the heart of the Einstein–Podolsky–Rosen paradox and Schrödinger analysis is the definition of an "element of reality," and the conclusion of the argument in each case is the existence of these physical properties. To reach this conclusion, as we are aware, one requires "prediction with certainty". In the scenario of the EPR experiments, such prediction becomes possible after one of the particles has undergone a measurement procedure to ascertain its value. Because the system is perfectly correlated, the second particle's value may be predicted with absolute conviction. Note well that nothing is said here regarding the quantum rules concerned with the form and evolution of the wave function.[46] Aside from[47] the perfect correlations themselves, all that we truly require to derive the Schrödinger paradox conclusion (and Schrödinger's Theorem) is that the values obtained in measurement procedures[48] are the eigenvalues or joint-eigenvalues of the associated operators. For brevity, we call this the "eigenvalue assumption." With this stipulation from the quantum formalism, we have sufficient grounds to make the implied noncontextual variables adhere to the correct quantities.[49]

Based upon the results just shown, it is obvious that Schrödinger's Theorem does not depend upon the full breadth of the quantum formalism, but only the eigenvalue assumption. With explicit reference to this assumption, we may restate the theorem: assuming locality, along with perfect correlations and the eigenvalue assumption, it

---

[46] See Sect. 1.3.

[47] Some might object, pointing to the fact that the detailed form of the wave function played an indispensable role in deriving the perfect correlations. While it is true that this served to draw our attention to the unusual physical situation, nevertheless, when constructing Schrödinger's Theorem, we may "forget" this fact and simply *begin* with the perfect correlations. In others words, it is the perfect correlations *in and of themselves* that are needed for the logic, such that (with locality) one is led to the conclusion of definite values on the observables. The matter of the origin of these correlations is orthogonal to the argument.

[48] Clearly, one must assume that the measurement procedures themselves may be performed.

[49] If one cared to entertain other speculations, one might wish to assume that the perfect correlations exist, but that the spin-squared components when measured all take the value, say, π, 0 and $-\pi$, rather than the quantum predictions of 1, 0 and -1. In that case, the Schrödinger paradox argument would lead us to conclude that the value map exists but takes the observables to the wrong values. Likewise, one could assume that the measurements offer results from the set 1, 0, -1, but that the rule regarding joint-eigenvalues of commuting sets S_x^2, S_y^2, S_z^2 is not obeyed. Under that assumption, the Schrödinger paradox would conclude hidden variables which violate the quantum rule regarding joint-eigenvalues.

follows that there exist noncontextual hidden variables on all observables of each particle.

Recall now Sect. 4.4.3, where we developed a maximally entangled state consisting of two spin-1 particles (see Eq. 4.42) and discussed the perfect correlations between the observables $\{S^2_{\theta, \phi}\}$. When we consider Schrödinger's Theorem in this case, it is clear that the eigenvalue assumption reduces to the SPIN axiom.

Now, we are prepared to bring together all that has been developed. For the maximally entangled quantum state (4.42), the perfect correlations that appear in the Schrödinger paradox are identical to the TWIN axiom. The locality assumption is the same as FIN. And finally we saw that the eigenvalue assumption is equivalent to SPIN. Thus, if we consider Schrödinger's Theorem for the quantum state (4.42), we have that SPIN, TWIN and FIN lead to the existence of a noncontextual value map on the spin-squared observables $\{S^2_{\theta, \phi}\}$ of the two spin-1 particles. This is exactly what is represented by the Conway and Kochen form $\theta_0(w)$ in Eq. 4.58.

Armed with this information, if we then return to the analysis given by Conway and Kochen, the situation looks quite different. Apart from making use of the Kochen and Specker theorem, the Free Will Theorem depends also on the following argument. Given SPIN, TWIN and FIN, any deterministic theory of quantum mechanics must reduce to a simple value map function $\theta_0(w)$ on the spin-squared observables in question, if we allow for human free will. Comparing this to Schrödinger's Theorem, Conway and Kochen have brought us to the same conclusion *but based upon five assumptions rather than three.*

It is in this sense that Conway and Kochen have overlooked the potency of SPIN, TWIN and FIN—assumptions powerful enough imply the existence of a value-map on the spin observables. The additional assumptions Conway and Kochen utilize— that of deterministic quantum formulation $\theta_a(x, y, z; \alpha')$, and that of human free will—are superfluous[50] to the conclusion they reach,[51] namely the noncontextual mapping $\theta_0(w)$. That the free will assumption is unneeded effectively decouples the logic of the argument from this issue. That the deterministic quantum formulation is unneeded decouples the argument from any impact on such constructions.

One might say that when Conway and Kochen arrive at the value map functions $\theta_0(w)$ in Eq. 4.58 they have found a "half truth." They are correct in that these

---

[50]   The EPR/Bell argument is sometimes mistakenly associated with free will, but in Sect. 3.4.3 we showed that all that is required (in addition to locality and perfect correlations) is the assumption of "no-superdeterminism" (by the same argument, one can show that Schrödinger's Theorem also depends on this axiom.) As we demonstrated, this assumption is weaker than the free will axiom. The reader should note well that the no-superdeterminism axiom is certainly *not unique* to EPR/Bell and the Schrödinger paradox. The Conway Kochen argument itself and indeed all scientific arguments must hold to this fundamental postulate. Some may believe that it would be better to reconstruct all arguments in this section in such a way that the no-superdeterminism assumption is explicitly mentioned throughout. However, we regard this assumption as so basic to the very meaning of scientific activity that frequent allusions to the principle would be quite redundant.

[51]   We do not not refer here to the conclusion of the Free Will Theorem itself, but that of what we call the SPIN–TWIN–FIN argument. We noted above that the Free Will Theorem consists of two components, the Kochen and Specker Theorem and the argument which makes direct use of SPIN, TWIN and FIN. We are discussing here only the latter component.

functions certainly follow at this stage of the argument. However, they are incorrect insofar as they imply that their existence depends upon[52] the initial assumption of deterministic quantum formulation given by functions $\theta(x =?, y, z; \alpha')$. As we have seen, all that is required are the three assumptions SPIN, TWIN and FIN.

The second part of the Free Will Theorem consists in combining these results with the Kochen and Specker theorem. As one does so, one finds that the conclusion of the former argument is proven impossible, in that the value-map must contradict the quantum requirements.

If one had mistakenly missed the power of the three axioms SPIN, TWIN and FIN, one might have traced this conflict back to either the free will assumption or the realistic quantum formulation represented by functions such as $\theta(x =?, y, z; \alpha')$. Since human free will is something we quite naturally wish to maintain, one might then turn to the theoretical formulation as the root of the problem.

However, we know from the Schrödinger paradox (and Schrödinger's Theorem) that the three axioms SPIN, TWIN and FIN in and of themselves suffice to lead one to the value map $\theta_0(w)$, and subsequently to contradiction with the quantum mechanical predictions via the Kochen and Specker theorem. As we emphasized in the discussion of Schrödinger nonlocality, the means that any hypothetical theory satisfying these requirements must conflict with the quantum predictions. SPIN and TWIN cannot lead to such a conflict, since they are themselves predictions of quantum theory. Therefore, there is no alternative but to conclude that the FIN axiom brings about the incongruity, and we again find ourselves concluding that quantum predictions conflict with all local theoretical explanations. Note that this argument differs[53] in no essential way from the nonlocality proof[54] presented in Sect. 4.5.1.

Moreover, one cannot appeal to the "Strong Free Will Theorem" [14] for an argument against determinism in quantum physics. The added strength of that theorem is due to the authors weakening their axioms related to free will and to the locality and independence of the two physical systems. Human free will and FIN are here replaced by the axiom "MIN," which consists of a weakened form of locality, as well as a weakened version of the free will assumption.

Let us first discuss the new form of locality. In fact, this change was briefly alluded to[55] in our presentation of Schrödinger's Theorem in Sect. 4.3.2.[56] As opposed to the

---

[52] Neither does this conclusion—the existence of such functions—depend upon the axiom of human free will.

[53] The appearance of the argument in this section is slightly different, as we explicitly bring out the dependence of Schrödinger's Theorem on the eigenvalue assumption (which reduced to TWIN in this case). Therefore, we have expressed Schrödinger's Theorem in a slightly different form here, so that the eigenvalue assumption joins the familiar premises of locality and perfect correlations.

[54] To be exact, one can create such a proof from *any* spectral incompatibility argument. However, the comparison we are drawing is to that which involved Kochen and Specker's theorem.

[55] Please see the discussion of locality on page 68, footnote 13.

[56] We discussed there the general case of correlated observables on an infinite dimensional state, rather than the particular one of the spin-1 observables present in the Conway Kochen theorem. The case of a system with perfect correlations in terms of spin-1 observables was given in Sects. 4.4.3 and 4.5.1.

basic concept of locality, according to which no information of any kind might be transmitted faster than light, all that is required for Schrödinger's Theorem is a more limited axiom. Schrödinger's Theorem depends only on the assertion that the choice of measurement procedure $\mathcal{M}(A)$ used to determine A cannot influence the result of measurement of its correlated partner $\tilde{A}$. We referred to this as "measurement locality" in the discussion on page 68. This is the same as the weakened form of locality used by Conway and Kochen.[57]

Conway and Kochen add to this weakened locality a stipulation regarding human free will. They write that the choices of experimenters A and B regarding which triplet x, y, z of spin components they measure are made freely and that the experimental results cannot depend on such choices. MIN consists[58] of the conjunction of measurement locality with this reduced form of free will. Note that measurement locality, or "ML", *is a more minimal assumption than MIN* since the former makes no mention of free will.

In our analysis of the original Free Will Theorem, we showed that the assumptions of a deterministic quantum formulation and that of human free will are unneeded. Without these superfluous axioms, the logic of the full argument finally assumed this form[59]: assuming SPIN, TWIN and FIN, we are led to a conflict with the quantum predictions. With the new attention given to weakening the locality condition, we find that the Schrödinger paradox can be strengthened by assuming only "measurement locality," or ML. That being the case, we have finally that the implications of the whole analysis are as follows. Assuming SPIN, TWIN and ML, one is led to a contradiction with quantum. From this, we can only find that quantum theory conflicts with ML, i.e., we once again conclude quantum nonlocality.

If both the original and strong version of the Free Will Theorem are simply proofs of quantum nonlocality, what is implied for quantum realism? The reasoning of Conway and Kochen essentially sees the two components of their proof as separate and distinct entities. Having long ago "disposed" of noncontextual hidden variables in the original Kochen and Specker theorem,[60] now Conway and Kochen ask us to accept that a new component might be added which defeats the possibility of contextual hidden variables. At the same time, they seem to be asserting that to seriously consider hidden variables, one must also accept the denial of human free will. However, their conclusion is based on several basic oversights, which we have pointed out. First, within the main component of their argument, i.e., that which is

---

[57] Conway and Kochen refer specifically to those measurements which arise in the Peres proof [15] of the Kochen and Specker theorem. But placing such a restriction upon measurement locality has no impact on these matters. Schrödinger's Theorem still follows. Also, they refine their description of locality, giving it in terms of the particles' past half-spaces, rather than past light cones. Such a change does not alter the basic form of the logic that follows from the Schrödinger paradox.

[58] Along with a refinement in the description of locality in terms of the particles' past half-spaces, rather than past light cones.

[59] By "full argument" we refer to the combination of what we called the SPIN–TWIN–FIN argument with the Kochen and Specker theorem. Such a combination forms a complex argument, as we have now discussed several times.

[60] Actually Gleason's theorem offers essentially the same result, as we have seen.

designed to supplement the destruction of noncontexual hidden variables with the defeat of contextual, they have overlooked that the axioms SPIN FIN and TWIN are more powerful than they know. These axioms imply the existence of noncontextual elements of physical reality, per the arguments of the Schrödinger paradox. Having missed this fact, Conway and Kochen are then unaware that the two components of their paper, when linked into a complex argument, tell us that locality conflicts with quantum mechanical predictions. As such, the Conway and Kochen paper is shown for what it is: not a limitation upon quantum realism, but a restriction upon quantum mechanics itself.

We have seen that the Free Will Theorem is none other than a special case of Schrödinger nonlocality. Because their analysis depends on such a structure, it is clear that one can also compare it to the Einstein–Podolsky–Rosen paradox, and claim that its logic runs parallel to that famous argument.

As such, one might regard Conway and Kochen's oversight as the repeat of a common misunderstanding of the classic EPR argument. In the words of Bell [16]:

> It is important to note that to the limited degree to which *determinism* plays a role in the EPR argument, it is not assumed but *inferred*... ... It is remarkably difficult to get this point across, that determinism is not a *presupposition* of the analysis.

The point was difficult for Bell to convey, and it would seem that some of today's theorists may also fail to see it.

4.6 Schrödinger's Paradox and Von Neumann's Theorem

Finally, we would like to make a few observations which are of historical interest. We now focus on the line of thought Schrödinger followed subsequent to his generalization of the Einstein–Podolsky–Rosen paradox. In addition, we consider the question of the possible consequences had he repeated the mistake made by a well-known contemporary, or had he anticipated any of several mathematical theorems that were developed somewhat later.

Having concluded the existence of definite values on the observables, Schrödinger considered the question of the type of relationships which might govern these values. As we discussed in Sect. 1.4.6, he was able to show that no such values can obey the same relationships that constrain the observables themselves. In particular, he observed that the relationship

$$H = p^2 + a^2 q^2, \qquad (4.59)$$

is not generally obeyed by the eigenvalues of the observables H, p^2, q^2, so that no value map $V(O)$ can satisfy this equation. From this it follows[61] immediately that there exists no value map which is *linear* on the observables. Thus we see that

---

[61] To show this, one need only note that the value of any observable $f(O)$ will be f of the value of O, where f is any mathematical function (see Sect. 1.4.6).

Schrödinger's argument essentially leads to the same conclusion regarding hidden variables as von Neumann's theorem. Schrödinger did not, however, consider this as proof of the impossibility of hidden variables.

Instead, the fact that the definite values of his EPR generalization must fail to obey such relationships prompted Schrödinger to consider the possibility that *no* relationship whatsoever serves to constrain them [8] (emphasis due to original author): "Should one now think that because we are so ignorant about the relations among the variable-values held ready in *one* system, that none exists, that far-ranging arbitrary combination can occur?" Note, however, that if the 'variable-values' obey no constraining relationship, each must be an independent parameter of the system. Thus, Schrödinger continues with the statement: "That would mean that a system of '*one* degree of freedom' would need not merely *two* numbers for adequately describing it, as in classical mechanics, but rather many more, perhaps infinitely many."

Recall the discussion of the hidden variables theory known as Bohmian mechanics given in Sect. 2.5.1. There we saw that the state description of a system is given in this theory by the wave function ψ and the system configuration $\mathbf{q}$. The mathematical form of ψ is the same as in the quantum formalism, i.e., ψ is a vector in the Hilbert space associated with the system. Consider a spinless particle constrained to move in one dimension. The Bohmian mechanics state description would consist of the wave function $\psi(x) \in L_2$ and position $x \in \mathbb{R}$. Since any L_2 function $\psi(x)$ is infinite dimensional, i.e. it is an assignment of numbers to the points $x \in (-\infty, \infty)$, the Bohmian mechanics state description is infinite-dimensional. Thus, a theory of hidden variables such as Bohmian mechanics provides just the type of description that Schrödinger's speculations had led him to conclude.

Let us now suppose that rather than reasoning as he did, Schrödinger instead committed the same error as von Neumann. In other words, we consider the possible consequences had Schrödinger regarded the failure of the definite values to satisfy the same relations as the observables as proof that no such values can possibly agree with quantum mechanics. This false result would appear to refute the conclusion of Schrödinger's generalization of EPR; it would seem to imply that the Schrödinger paradox leads to a conflict with the quantum theory. The Schrödinger paradox, like the EPR paradox, assumes only that the perfect correlations can be explained in terms of a local theoretical model. Hence, if Schrödinger had concluded with von Neumann that hidden variables must conflict with quantum mechanics, then he would have been led to deduce quantum nonlocality.

As we saw above, such a conclusion actually follows from the combination of Schrödinger's paradox with any of the spectral incompatibility theorems, for example, those of Gleason, and Kochen and Specker. Since Gleason's theorem involves the proof of the trace relation (2.3), it seems reasonable to regard it as being the most similar of these theorems to that of von Neumann, which features a derivation of this same formula.[62] Note that to whatever degree one might regard Gleason's theorem as being similar to von Neumann's, one must regard Schrödinger as having come just that close to a proof of quantum nonlocality.

---

[62] Von Neumann's derivation is based on quite different assumptions, as we saw.

4.7 Summary and Conclusions

The investigation presented in this book has been aimed towards illuminating the issue of realism in quantum physics, most especially as it is tied to Bell's theorem. We approached the problem by first considering other famous mathematical theorems concerned with the possibility of forumulating an objective quantum theory. We saw that the early effort of John von Neumann as well as those of Gleason and of Kochen and Specker all fall short due to various misinterpretations of the quantum measurement process. Bell's theorem is somewhat more subtle in its implications: taken on its own, one might imagine that this argument succeeds where von Neumann and the others failed. However, as Bell and others showed, this is simply not the case: the theorem's conclusions leave us with the irreducible nonlocality of quantum mechanics. As such, Bell's famous analysis does not offer special limitations applying only to realistic formulations, but a general result which constrains quantum theory itself as well as all possible interpretations.

The representation of a hypothetical formulation of quantum physics is a task which requires careful attention to the quantum formalism and its rules. The major flaw in the "no-hidden-variables proofs" is that they neglect the detailed structure of the quantum measurement procedures. Von Neumann's argument failed in a some-what trivial way, as it disregards the fundamental directive that incompatible observables cannot be measured simultaneously. Both Gleason's, as well as Kochen and Specker's theorem, may seem at first to succeed where von Neumann failed, i.e., to prove the impossibility of hidden variables. Nevertheless, as Bell has shown [17, 18], these theorems also fail as arguments against hidden variables, since they do not account for contextuality. This concept is easily illustrated by examination of the quantum formalism's rules of measurement. We find that the 'measurement of an observable O' can be performed using distinct experimental procedures $\mathcal{E}(O)$ and $\mathcal{E}'(O)$. That $\mathcal{E}$ and $\mathcal{E}'$ are distinct is especially obvious if these measure the commuting sets $\mathcal{C}$ and $\mathcal{C}'$ where $\mathcal{C}$ and $\mathcal{C}'$ both contain O, but the members of $\mathcal{C}$ fail to commute with those of $\mathcal{C}'$. It is therefore quite reasonable to expect that a hidden variables theory should allow for the possibility that different procedures for measurement of some observable might yield different results.

Examples of just such incompatible commuting sets are found among the observables in each of the theorems we addressed in Chap. 2: Gleason's and Kochen and Specker's. For example, in the theorem of Kochen and Specker, the commuting sets are simply of the form $\{S_x^2, S_y^2, S_z^2\}$, i.e., the squares of the spin components of a spin 1 particle taken with respect to some Cartesian axis system x, y, z. Here one can see that a given observable S_x^2 belongs to both $\{S_x^2, S_y^2, S_z^2\}$, and $\{S_x^2, S_{y'}^2, S_{z'}^2\}$, where the y' and z' axes are oblique relative to the y, and z axes. Since the theorems of Gleason, and Kochen and Specker consider a function $E(O)$ which *assigns a single value to each observable*, they cannot account for the possibility of incompatible measurement procedures which the quantum formalism's rules of measurement allows us. Clearly, the approach taken by these theorems falls short of addressing the hidden variables issue properly. Thus, we come to concur with J.S. Bell's assessment that [18] *"What is proved by impossibility proofs . . . is lack of imagination."*

To address the general question of hidden variables, one must allow for this important feature of contextuality. Gleason's theorem, and also Kochen and Specker's may be seen as proofs that this simple and natural feature is necessary in any hidden variables theory. In particular, any attempt to construct a value map that neglects this feature will fail in that it cannot satisfy the relationships constraining the commuting sets. We have seen that there is a simpler way to express such results: there exists no value map on the observables which carries every commuting set to one of its joint-eigenvalues. We referred to this as "spectral incompatibility."

To evoke the property of contextuality in quantum physics recalls the view of Niels Bohr, who warns us of [19, p. 210] "the impossibility of any sharp separation between the behavior of atomic objects and the interaction with the measuring instruments which serve to define the conditions under which the phenomena appear." Far from something contrived or artificial, contextuality in hidden variables is quite natural since it reflects the possibility of distinct experimental procedures for measurement of a single observable. This is nothing esoteric, but is a basic feature of the quantum formalism itself. As Bell expresses it [17]: "The result of an observation may reasonably depend not only on the state of the system (including hidden variables) but also on the complete disposition of the apparatus."

When we come to consider Bell's theorem, we must do so in the context of the Einstein–Podolsky–Rosen paradox.[63] The latter is a powerful result and we noted that it implies the following theorem when expressed in terms of Bohm's spin-singlet formulation.[64] Assuming locality and the perfect correlations on the spin-singlet state, it follows that there must exist noncontextual hidden variables on all spin components of each particle. Because the quantum mechanical description of the state does not account for such values, one must conclude, as EPR did, that this theory is incomplete. Bell's theorem essentially begins where the spin singlet EPR analysis concluded. In his analysis, Bell considers the statistical correlation between spin component measuring experiments carried out on the two particles. According to his famous theorem, the theoretical prediction for this correlation derivable from the variables $A(\lambda, \hat{a})$ and $B(\lambda, \hat{b})$ must satisfy 'Bell's inequality'. The prediction given by quantum mechanics does *not* generally agree with this inequality. Bell's theorem in itself provides a proof that local hidden variables must conflict with quantum mechanics.

It is important to note that what might *appear* to be Bell's assumption—the existence of definite values for all spin components—is identical to the conclusion of the spin-singlet version of the Einstein–Podolsky–Rosen argument. In fact, Bell assumes nothing beyond what follows from EPR. Therefore the proper way to assess the implications of these arguments is to combine them into a single analysis. According to logical terminology, one refers to such a conjunction as a "complex argument."

---

[63] Actually, the spin-singlet version thereof.

[64] In this chapter, we saw that analogous conclusions followed for the original EPR quantum state, as well as for every maximally entangled state.

In this case, the complex argument begins with the assumptions of the spin-singlet EPR paradox, and end with the conclusion of Bell's theorem. Therefore we have that the assumption of locality with perfect correlations leads to a conflict with the quantum predictions. In the words Bell himself [20]: *"It now seems that the non-locality is deeply rooted in quantum mechanics itself and will persist in any completion."*

According to the above results, the fact that nonlocality is required of hidden variables does not in any way diminish the prospect of such theories. Rather, what we have seen is that the true implications of such an analysis as Bell's is that quantum mechanics itself is irreducibly nonlocal.

In the fourth and final chapter, we addressed Erwin Schrödinger's generalization of the EPR paradox. Besides greatly extending the incompleteness argument of EPR, Schrödinger's analysis provides for a new set of "nonlocality without inequalities" proofs, which have several important features. The Schrödinger paradox concerns the perfect correlations exhibited not just by a single quantum state, but for a general class of states called the maximally entangled states. A maximally entangled state is any state of the form

$$\sum_{n=1}^{N} |\phi_n\rangle \otimes |\psi_n\rangle, \tag{4.60}$$

where $\{|\phi_n\rangle\}$ and $\{|\psi_n\rangle\}$ are bases of the (N-dimensional) Hilbert spaces of subsystems 1 and 2, respectively. As in the EPR analysis (both the spin singlet and the original version) Schrödinger gives an incompleteness argument, according to which there must exist definite values for all observables of both subsystems. It may be shown using Gleason's theorem, Kochen and Specker's theorem, or any other 'spectral incompatibility' proof, that such assigned values must *conflict* with the empirical predictions of quantum mechanics. Such a disagreement leads inevitably to quantum nonlocality, and we suggested results of this kind should be named "Schrödinger nonlocality" proofs.

The heart of such proofs is the conflict between the implications of Schrödinger's paradox and spectral incompatibility. We noted that such a proof differs from that developed within Bell's theorem, in that it involves predictions for individual measurements, rather than the statistics of a series. Thus, the Schrödinger nonlocality results are 'nonlocality without inequalities' proofs. Moreover, as we observed in Sect. 4.5.1, the required empirical disagreement arises in terms of the observables of *just one* of the two correlated subsystems.

The Conway and Kochen Free Will Theorem is a very recent paper purporting to be a logical proof that determinism can never be reestablished as the basis of quantum physics. The authors stress that the strength of their result follows from its simple and natural assumptions. Beyond this, they find themselves introducing human free will as an assumption to leverage the result that nature is irreducibly random. This indeed is an appealing picture to those without an intimate familiarity with the EPR paradox and its generalization by Erwin Schrödinger. On the other hand, with the aid of the latter one finds that the logical argument they offer does not justify their claims.

The linkage they make to human free will is quite superfluous, since the assumption of free will can be dropped or even reversed in the argument without affecting its conclusion. The argument's assumptions SPIN, TWIN and FIN[65] are just as reasonable as the authors claim, but more powerful than they appear to have realized. By the logical implications of the Schrödinger paradox (and Schrödinger's Theorem) these axioms are precisely what is needed to derive the inevitable existence of non-contextual hidden variables on the appropriate components of $S^2_{\theta, \phi}$. The fact that the Kochen and Specker theorem contradicts this leads finally to the conclusion that any theory satisfying FIN must conflict with the quantum predictions.[66] This is just the same as the Schrödinger nonlocality argument we gave in Sect. 4.5.1, in which we showed how a complex argument may be formed based on the Schrödinger paradox and the Kochen and Specker theorem. What Conway and Kochen have developed is a special case of the Schrödinger nonlocality proof.

All this might bring us to inquire how Schrödinger himself regarded his results, and what further conclusions he drew within his remarkable paper. Clearly, it would have been possible for him to argue for quantum nonlocality had he anticipated the results of any of a wide variety of theorems including at least Gleason's, Kochen and Specker's, or any other spectral incompatibility theorem. Instead, as we saw in Chap. 1, Schrödinger essentially reproduced the von Neumann argument against hidden variables, in his observation of a set of observables that obey a linear relationship not satisfied by the set's eigenvalues. What may be seen from von Neumann's result is just what Schrödinger noted—relationships constraining the observables do not necessarily constrain their values. Schrödinger continued this line of thought by speculating on the case for which *no* relation whatsoever constrained the values of the various observables. In light of his generalization of the EPR paradox, these notions led Schrödinger to the idea that the quantum system in question might possess an infinite number of degrees of freedom, which concept is actually quite similar to that of Bohmian mechanics.

If, on the other hand, Schrödinger *had* made von Neumann's error, i.e., had concluded the impossibility of a map from observables to values, this mistaken line of reasoning would have allowed him to reach the conclusion of quantum nonlocality. Thus, insofar as one might regard the von Neumann proof as "almost" leading to the type of conclusion that follows from Gleason's theorem, one must consider Schrödinger as having come precisely that close to a proof of quantum nonlocality.

It is quite interesting to see that many of the issues related to quantum mechanical incompleteness and hidden variables were addressed in the 1935 work [8–10] of Erwin Schrödinger. One could make the case that Schrödinger's work is the most far-reaching of the early analyses addressed to the subject. Schrödinger presented here not only his famous "cat paradox," but he developed results beyond those of the Einstein, Podolsky, Rosen paper—an extension of their incompleteness argument,

---

[65] As we noted in Sect. 4.5.2 there is nothing gained by appealing to the strong free will theorem in which FIN is replaced by MIN.

[66] Conflict with quantum mechanics cannot be laid at the feet of TWIN nor SPIN, since these are taken from the theory itself.

and an analysis of this result in terms closely related to von Neumann's theorem. It seems clear that the field of foundations of quantum mechanics might have been greatly advanced had these features of Schrödinger's paper been more widely appreciated at the time it was first published.

References

1. Brown, H.R., Svetlichny, G.: Nonlocality and Gleason's Lemma. Part I Deterministic Theories. Found. Phys. **20**, 1379 (1990)
2. Heywood, P., Redhead, M.L.G.: Nonlocality and the Kochen-Specker Paradox. Found. Phys. **13**, 481 (1983)
3. Aravind, P.K.: Bell's theorem without inequalities and only two distant observers. Found. Phys. Lett. **15**, 399–405 (2002)
4. Cabello, A.: Bell's theorem without inequalities and only two distant observers. Phys. Rev. Lett. **86**, 1911–1914 (2001)
5. Bassi, A., Ghirardi, G.C.: The Conway-Kochen Argument and relativistic GRW Models. Found. Phys. **37**, 169 (2007). Physics archives: arXiv:quantph/610209
6. Tumulka, R.: Comment on 'The Free Will Theorem'. Found. Phys. **37**, 186–197 (2007). Physics archives: arXic:quant-ph/0611283
7. Goldstein, S., Tausk, D., Tumulka, R., Zanghí, N.: What does the free will theorem actually prove? Notices AMS **57**(11):1451–1453 (2010) arXiv:0905.4641v1 [quant-ph]
8. Schrödinger, E.: Die gegenwA rtige Situation in der Quantenmechanik. Naturwissenschaften **23** 807–812, 823–828, 844–849 (1935). The English translation of this work - 'The present situation in quantum mechanics' - appears. In: Proceedings of the American Philosophical Society **124**, 323–338 (1980) (translated by J. Drimmer), and can also be found in [21, p. 152]
9. Schrödinger, E.: Discussion of probability relations between separated systems. Proc. Cambridge Phil. Soc. **31**, 555 (1935)
10. Schrödinger, E.: Probability relations between separated systems. Proc. Cambridge Phil. Soc. **32**, 446 (1936)
11. Bohm, D.: Quantum Theory. Prentice Hall, Englewood Cliffs (1951)
12. Einstein, A., Podolsky, B., Rosen, N.: Con quantum mechanical description of physical reality be considered complete? Phys. Rev. **47**, 777 (1935). Reprinted in [21, p. 138]
13. Conway, J., Kochen, S.: "The Free Will Theorem" Found. Phys. **36**(10); 1441–1473 (2006). Physics archives arXiv:quant-ph/0604079
14. Conway, J., Kochen, S.: The strong Free Will Theorem. Notices AMS **56**(2), 226–232 Feb 2009
15. Peres, A.: Quantum Theory: Concepts and Methods. Springer, Heidelberg (1995)
16. Bell, J.S.: Bertlemann's socks and the nature of reality Journal de Physique. Colloque C2, suppl. au numero 3, Tome 42 1981 pp. C2 41–61. Reprinted in [22, p. 139]
17. Bell, J.S.: On the problem of hidden variables in quantum mechanics. Rev. Modern Phys. **38** 447–452 (1966). Reprinted in [22, p. 1] and [21, p. 397]
18. Bell, J.S.: On the impossible pilot wave. Found. Phys. **12** 989–999 (1982) Reprinted in [22, p. 159]
19. Schilpp, P.A.: Albert Einstein: Philosopher-Scientist. Harper and Row, New York (1949)
20. Bell, J.S.: Quantum mechanics for cosmologists. p. 611 in: Quantum Gravity. vol. 2 Isham, C., Penrose, R., Sciama, D. (eds.) Clarendon Press, Oxford (1981). Reprinted in [22, p. 117]